Investigating The Potential Of Crispr Interference (CRISPRi) For Gene Regulation

Rickbed Nandi

Table of Contents

Chapter 1: Introduction to CRISPR Interference (CRISPRi)

1.1 What is CRISPRi?

In the sprawling scenery of genetic engineering, one tool has revolutionized the way we approach gene regulation: CRISPR Interference, or CRISPRi. The story of CRISPRi, and its remarkable journey from discovery to implementation, is a tale of precision and promise that has captured the imagination of researchers across various scientific disciplines. In this chapter, we embark on a journey of discovery ourselves, diving into the essence of CRISPRi, its fundamental mechanisms, and its profound implications in the intricate web of gene regulation.

The CRISPR Revolution

The term "CRISPR" first entered scientific discourse around the turn of the millennium. Initially, it was recognized as a peculiar, repetitive pattern found in the genomes of bacteria and archaea. Yet, it wasn't until the last decade that CRISPR, or Clustered Regularly Interspaced Short Palindromic Repeats, seized the scientific community's attention as a groundbreaking gene-editing tool.

At the core of this CRISPR system lies the CRISPR-associated (Cas) proteins, which serve as molecular scissors capable of cutting and manipulating DNA with extraordinary precision. However, the focus of this chapter is not on gene editing, but rather on a complementary and versatile function of CRISPR, known as CRISPR Interference, or CRISPRi.

CRISPRi: A Molecular Dimmer Switch

While CRISPR-Cas9, often associated with genome editing, cleaves DNA at precise locations to alter the genetic code, CRISPRi takes a different approach. CRISPRi acts as a gene regulator, a dimmer switch for the expression of specific genes. It doesn't directly modify the DNA sequence; instead, it modulates the degree to which a particular gene is transcribed into RNA, and subsequently, translated into proteins.

At the heart of CRISPRi is a non-cutting mutant of the Cas9 protein, called dCas9. This dCas9 variant retains its ability to bind to specific DNA sequences but lacks the ability to cleave them. Coupled with a guide RNA molecule (gRNA) that directs dCas9 to the target gene's promoter region, CRISPRi allows researchers to precisely control the extent to which a gene is turned on or off. This level of specificity is nothing short of revolutionary.

Unpacking the CRISPRi Mechanism

The mechanism by which CRISPRi operates is as elegant as it is effective. When dCas9, guided by its associated gRNA, attaches itself to a gene's promoter region, it obstructs the RNA polymerase enzyme's access. RNA polymerase is a crucial player in the transcription process, as it's responsible for copying DNA into RNA. By interfering with this transcription machinery, CRISPRi effectively halts or hinders the gene's expression.

Here's an analogy to help illustrate the concept: Imagine dCas9 as a molecular traffic cop, directing traffic at a busy intersection. The gRNA is like a map, telling the traffic cop which cars (genes) to stop or slow down (suppress). As the traffic cop stands in the middle of the road (the gene's promoter region), vehicles (RNA polymerase) heading towards their destination (gene expression) are directed to pause or detour. In this way, CRISPRi fine-tunes gene expression with remarkable precision.

Broad Applications of CRISPRi

The applications of CRISPRi are as diverse as the genes it can regulate. From fundamental research to therapeutic development, CRISPRi has opened doors to myriad possibilities. For instance, in basic research, scientists use CRISPRi to explore gene functions by selectively turning genes on or off. This helps us understand the intricate molecular mechanisms that govern biological processes.

In the realm of therapeutics, CRISPRi offers a potential way to treat genetic diseases by modulating the expression of disease-causing genes, without permanently altering the underlying DNA. This is particularly promising in cases where traditional gene-editing approaches, like CRISPR-Cas9, might carry unacceptable risks.

Moreover, CRISPRi has found applications in agriculture, offering the potential to engineer crops for enhanced yield, resistance to pests, and improved nutritional content. In synthetic biology, it enables the construction of finely-tuned genetic circuits for various biotechnological purposes, ranging from biofuel production to bioremediation.

The Road Ahead

As we navigate the landscape of CRISPRi, it's essential to recognize that this technology is not without its challenges. Ethical considerations, off-target effects, and the need for further optimization are among the hurdles that researchers are actively addressing. Nevertheless, CRISPRi's journey from its discovery to its far-reaching applications is a testament to the enduring spirit of scientific exploration and innovation.

In the chapters that follow, we will delve deeper into the molecular intricacies of CRISPRi, explore its applications across diverse fields, and consider the ethical and regulatory dimensions of this revolutionary gene regulation tool. Join us in this fascinating journey, where the molecular scalpel of CRISPRi opens doors to unprecedented insights into the genetic wall-hanging of life.

Therefore, CRISPRi is a versatile and precise gene regulation tool, offering researchers the means to control gene expression without altering the underlying DNA sequence. With the dCas9 protein and gRNA guiding its actions, CRISPRi has myriad applications in basic research, therapeutics, agriculture, synthetic biology, and more. This chapter has introduced us to the essence of CRISPRi, and in the subsequent sections, we will explore its applications and implications in greater detail.

1.2 Historical Overview of CRISPR Technology

The history of CRISPR technology is a fascinating journey that has fundamentally altered the course of molecular biology, genetic engineering, and biotechnology. It's a tale of relentless curiosity, collaboration, and innovation. In this section, we will embark on a journey through time to explore the key milestones in the development of CRISPR technology, tracing its evolution from a mere curiosity to the revolutionary gene-editing tool we know today.

Early Pioneers: The Curious Beginnings

Our story begins not in a state-of-the-art laboratory, but rather with a group of scientists who were simply trying to understand a peculiar aspect of the bacterial immune system. In the late 1980s, Japanese scientist Yoshizumi Ishino unwittingly stumbled upon a series of repeating DNA sequences while studying a strain of Escherichia coli, commonly known as E. coli. These sequences puzzled Ishino and his team, but it was not until 2000 that they would be coined as Clustered Regularly Interspaced Short Palindromic Repeats or, as we fondly call them, CRISPR. At this point, these sequences were an enigma in search of a purpose.

While Ishino's team in Japan was delving into the mystery of CRISPR sequences, another set of researchers, this time in Spain, were also intrigued by these strange patterns. Francisco Mojica, a microbiologist, was conducting his own investigations on extremophiles, bacteria that thrive in extreme conditions. Mojica's work with a specific extremophile, Haloferax mediterranei, led him to the conclusion that these odd sequences were likely linked to an immune response in bacteria, though the full picture remained elusive.

CRISPR as an Adaptive Immune System

The first hints of CRISPR's role as an adaptive immune system for bacteria were starting to emerge. Fast forward to 2005 when, inspired by Mojica's work, a group of researchers at the University of California, Berkeley, began examining the role of CRISPR in the bacterium Streptococcus thermophilus, used in yogurt production. They uncovered evidence suggesting that these CRISPR sequences played a role in defending bacteria against viruses. The idea that bacteria possessed a sophisticated, adaptable defence system against viral invaders was indeed a groundbreaking revelation.

The pieces of the CRISPR puzzle were coming together. It became clear that bacteria had a memory of past viral attacks, stored as segments in the CRISPR array. These segments were the equivalent of genetic mug shots, enabling the bacteria to recognize and fend off familiar viral assailants. But an even more remarkable aspect of CRISPR's story was yet to unfold.

CRISPR-Cas9: The Game Changer

The turning point in the CRISPR saga came in the form of the discovery of a molecular component that would later transform CRISPR from an intriguing biological phenomenon into a powerful gene-editing tool. In 2012, Jennifer Doudna and Emmanuelle Charpentier, working in collaboration, made a seismic breakthrough. They identified a specific protein within Streptococcus pyogenes bacteria called Cas9. This protein had a unique ability: it could precisely cut DNA. Moreover, the researchers found that by providing a synthetic RNA guide, they could redirect Cas9 to target specific DNA sequences, effectively creating a molecular scalpel for genetic material.

The implications of this discovery were profound. Suddenly, the scientific community had a tool that allowed them to selectively modify genes with unparalleled precision. The era of CRISPR-Cas9 had officially begun.

Rapid Adoption and Refinement

As word of this revolutionary technique spread, laboratories worldwide started experimenting with CRISPR-Cas9. It was quickly adopted as the method of choice for gene editing due to its simplicity and efficiency. Scientists found that they could use CRISPR-Cas9 to not only disable genes but also introduce new genetic material.

In 2013, George Church and his team at Harvard Medical School expanded the CRISPR toolbox. They harnessed the power of CRISPR for a variety of applications, including the editing of genes in human cells. Meanwhile, Feng Zhang, from the Broad Institute of MIT and Harvard, and his group were simultaneously developing CRISPR-Cas9 for use in eukaryotic cells. These parallel advances intensified the race to unlock the full potential of CRISPR.

Ethical and Regulatory Challenges

The rapid ascent of CRISPR technology was met with growing concerns regarding ethics and regulations. As the possibility of editing the human genome loomed larger, the international scientific community was faced with complex questions about the moral and legal implications of CRISPR's capabilities.

One event that heightened these concerns was the revelation by Chinese scientist He Jiankui in 2018, that he had edited the genes of twin girls to make them resistant to HIV. The international backlash against such unsanctioned human gene editing experiments underscored the urgency of establishing guidelines and regulations to govern the use of CRISPR technology in humans.

Expanding Horizons

The story of CRISPR technology does not end with Cas9. Researchers have continued to refine and expand its applications. New versions of CRISPR, such as CRISPR-Cas12 and CRISPR-Cas13, have emerged, each with unique features and potential applications. These variants have extended the gene-editing toolkit, making it more versatile and adaptable to different genetic tasks.

Beyond human health, CRISPR technology has found applications in agriculture, with the potential to create genetically modified crops that are more resistant to pests, diseases, and environmental stressors. It has also been instrumental in conservation efforts, helping to protect endangered species and restore damaged ecosystems.

The Future of CRISPR

The trajectory of CRISPR technology from its humble beginnings as mysterious bacterial sequences to its current status as a transformative gene-editing tool demonstrates the remarkable power of scientific discovery and innovation. However, the journey is far from over. The future of CRISPR promises even greater potential, with ongoing research into improving precision, reducing off-target effects, and broadening its scope of applications.

As the story of CRISPR unfolds, it is essential to navigate the complex ethical, legal, and social landscapes that this technology presents. With responsible use and continued advancements, CRISPR holds the potential to revolutionize medicine, agriculture, and biotechnology in ways that were once unimaginable. The history of CRISPR is an ongoing narrative, and the chapters yet to be written will undoubtedly shape the future of science and humanity.

1.3 The Significance of Gene Regulation

The dynamic interplay of genetic information, fundamental to life, lies at the heart of every living organism. Our very existence is an intricate dance of genes choreographing the processes that shape us. The story of life, though woven into the code of DNA, is not a fixed narrative but an unfolding masterpiece, with each gene a character playing a unique role. In the symphony of biology, gene regulation holds the conductor's baton, dictating when, where, and how each gene should contribute to the performance. This chapter delves into the exquisite world of gene regulation, and why it is so vital to our understanding of life and the promise of CRISPR Interference (CRISPRi).

Gene regulation is not a niche concept, limited to geneticists or bioinformaticians. It is a fundamental biological process that underpins our very existence, from the simplest microbe to the most complex human being. Gene regulation is the mechanism by which our cells control when and to what extent genes are active. If we liken the DNA to an extensive library of books, gene regulation decides which books are open, when they are read, and how much of the content is reviewed.

Imagine, for a moment, the human body as a city, and genes as the different buildings. Each gene, like a building, has its own unique function, which contributes to the overall life of the city. In this city, the regulation of gene activity functions as the city's governing body, determining when and how these buildings are used, ensuring that the city functions harmoniously. This governing body can adapt to various circumstances, directing resources to hospitals during a health crisis or shifting focus to schools when the population is young and growing. This dynamic adaptation and fine-tuning of gene activity are critical for our health, development, and survival.

Gene Regulation in Development

One of the most striking examples of gene regulation's significance is embryonic development. In the early stages of life, all the genetic information needed to create a human being is contained within a single cell. As this cell divides and differentiates into various cell types, the importance of precise gene regulation becomes evident.

For instance, consider the human heart. During development, a complex network of genes orchestrates the formation and function of this vital organ. These genes are activated and silenced at precise moments, ensuring that the heart forms correctly, contracts rhythmically, and pumps blood efficiently. Disruptions in this process can lead to congenital heart defects, which affect thousands of infants worldwide. Gene regulation is not just about turning genes on and off; it's about orchestrating a symphony that creates and maintains life.

Maintaining Homeostasis

Gene regulation is not limited to development. Throughout life, it plays a crucial role in maintaining the body's equilibrium. Our bodies are remarkably adept at adapting to changes in the environment, and this adaptability is largely driven by the dynamic nature of gene regulation.

For example, consider the regulation of blood sugar levels. After a meal, blood sugar levels rise, triggering a cascade of gene regulation events. Some genes promote the uptake of glucose into cells, while others stimulate the liver to store excess glucose as glycogen. This orchestrated response ensures that blood sugar levels return to normal, preventing hyperglycaemia. In contrast, when blood sugar levels drop, a different set of genes is activated to release stored glucose from the liver into the bloodstream. This intricate balancing act is vital for our health, as it prevents conditions like diabetes.

Disease and Gene Regulation

Conversely, malfunctioning gene regulation can have severe consequences for health. Many diseases, from cancer to neurological disorders, can be attributed to aberrant gene regulation. In cancer, for instance, genes that should inhibit cell growth are silenced, while genes promoting uncontrolled growth are activated. Understanding and modulating gene regulation is at the forefront of research aimed at finding effective treatments for these diseases.

CRISPRi: A Revolutionary Tool for Gene Regulation

Given the pivotal role of gene regulation in health and disease, the development of tools like CRISPRi is revolutionary. CRISPRi empowers researchers to manipulate gene activity with unprecedented precision. It is akin to having a "genetic dimmer switch" that can fine-tune gene expression levels.

Moreover, CRISPRi is not confined to laboratory research; it holds immense potential for therapeutic applications. As we journey further into this book, we will explore how CRISPRi is being harnessed to correct gene dysregulation, opening doors to innovative treatments for various diseases. It also offers new avenues for personalized medicine, tailoring therapies to individuals' unique genetic profiles.

Thus, gene regulation is the conductor of the genetic symphony that is life. It shapes our development, maintains our health, and is central to the emergence of disease. The advent of CRISPRi provides a powerful tool for understanding, manipulating, and ultimately harnessing the intricate world of gene regulation. As we embark on this exploration, we will uncover the remarkable potential of CRISPRi and its profound implications for science, medicine, and the future of biology.

Chapter 2: Molecular Mechanisms of CRISPRi

2.1 CRISPRi Components: dCas9 and Guide RNA

Gene regulation lies at the heart of cellular and molecular biology, and its precise control is crucial for maintaining normal cellular functions. In this chapter, we embark on a journey into the remarkable world of CRISPR interference, or CRISPRi, an innovative and powerful tool for gene regulation that has reshaped the field of molecular biology. Central to this technique are the two pivotal components - the deactivated Cas9 (dCas9) protein and the guide RNA (gRNA), working in perfect harmony to exercise exquisite control over gene expression.

dCas9: The Molecular Sentinel

At the heart of CRISPRi is the dCas9 protein, a mutation of the famed Cas9 protein originally identified in the CRISPR-Cas system of Streptococcus pyogenes. In its natural form, Cas9 is a DNA-cleaving enzyme that acts like a pair of molecular scissors, making double-stranded cuts in DNA, serving as the go-to component in the more popular CRISPR-Cas9 genome editing system. However, in CRISPRi, dCas9 plays an altogether different role. Instead of being the molecular scissor, it functions as a sentinel, a guard post at the gates of the genetic code, preventing the transcription of specific genes.

The critical feature of dCas9 that makes it an ideal candidate for this purpose is its inactivation. Scientists have ingeniously harnessed a point mutation in the Cas9 protein, rendering it catalytically inactive. This means that dCas9 is incapable of performing its usual DNA-cleaving role, leaving it as a precise and controllable DNA-binding protein.

The dCas9 protein serves as the guardian of gene expression, attaching itself to specific DNA sequences like a lock and key. This lock-and-key interaction is achieved through base pairing - where the guide RNA (gRNA) takes centre stage. However, it is important to note that dCas9 does not act alone; it requires a partner in the form of the gRNA to accurately locate and target specific genes for regulation.

Guide RNA (gRNA): The GPS of CRISPRi

The guide RNA, or gRNA, is the navigational system of CRISPRi. It is a short synthetic RNA molecule designed to recognize and bind to a specific DNA sequence of interest. This sequence is typically found within the promoter region of a target gene, where transcription initiation occurs. The gRNA is engineered with the utmost precision to create a complementary match to the target DNA sequence.

Crucially, the gRNA does more than just guide dCas9 to the right location in the genome; it also carries specific modifications that enhance its performance. The 5' end of the gRNA contains a sequence recognized by the Cas9 protein, ensuring that the two components can bind together effectively. Additionally, the 3' end of the gRNA is designed to prevent the initiation of transcription. This acts as a roadblock to RNA polymerase, the enzyme responsible for transcribing the gene. When the gRNA binds to its target DNA sequence, it brings dCas9 along with it, effectively barricading the gene from transcription and halting the synthesis of messenger RNA.

The incredible specificity of the gRNA is one of the chief strengths of CRISPRi. By carefully engineering the gRNA, researchers can dictate with remarkable precision which genes are to be silenced, leaving neighbouring genes unaffected. This level of fine-tuned control has opened up a world of possibilities in gene regulation, as scientists can now selectively modulate the expression of individual genes in the vast genomic landscape.

Synergy Between dCas9 and gRNA

The synergy between dCas9 and gRNA is what makes CRISPRi such a powerful and versatile tool for gene regulation. The dCas9 protein, with its ability to latch onto specific DNA sequences without altering them, acts as the anchor, while the gRNA provides the guidance. Together, they create a lock-and-key mechanism that offers exquisite control over gene expression.

The process unfolds as follows: the gRNA, tailored to a particular gene of interest, attaches to the dCas9 protein, forming a complex. This complex then scans the genomic terrain, searching for the designated DNA sequence. Once it finds its target, the gRNA-dCas9 complex locks onto the gene's promoter region. The dCas9 component binds to the DNA like a hand in a glove, while the gRNA acts as the guardrail preventing the transcription machinery from approaching the gene.

This combination of dCas9 and gRNA forms a powerful genetic blockade. In essence, it places the gene under house arrest, inhibiting its expression without modifying its DNA sequence. It's a bit like controlling the flow of traffic in a busy city, with dCas9 and gRNA functioning as traffic police directing and redirecting vehicles to ensure a smooth and controlled flow.

Versatility and Fine-Tuning

One of the remarkable features of CRISPRi is its versatility and fine-tuning capabilities. By simply designing a different gRNA, researchers can target a multitude of genes with precision. This flexibility makes CRISPRi an invaluable tool for both basic research and potential therapeutic applications.

For instance, if a researcher is interested in studying the effects of silencing a particular oncogene, they can design a gRNA to target that gene specifically. Alternatively, they can use CRISPRi to investigate the function of a gene in a complex biological pathway. The ability to turn genes on and off at will provides invaluable insights into gene function and regulation.

Furthermore, the degree of gene silencing can be finely adjusted by altering the concentration of the dCas9-gRNA complex. This allows for the modulation of gene expression from a complete shutdown to partial inhibition, providing a spectrum of control that fits the specific requirements of the experiment or application.

In short, the two fundamental components of CRISPRi, dCas9 and gRNA, form a dynamic duo in the realm of gene regulation. dCas9, with its inactivated cutting ability, acts as the molecular sentinel, while the guide RNA is the GPS, steering the complex toward its target. Together, they create a versatile and precise tool that is revolutionizing our understanding of gene regulation and offering exciting prospects for diverse applications in the biological and medical sciences.

2.2 Targeting Specific Gene Expression

The fundamental power of CRISPR interference (CRISPRi) lies in its ability to precisely and selectively target specific genes. In this section, we will explore the mechanisms that allow researchers to pinpoint and modulate individual gene expression. This precision is like a skilled archer hitting a bullseye, ensuring that the gene of interest is influenced while leaving the rest of the genomic landscape untouched.

A Guided Missile: The Role of Guide RNAs

To achieve the precision of CRISPRi, one must first understand the role of guide RNAs (gRNAs). These molecules act as the navigators for the CRISPR system. Imagine them as GPS coordinates directing the molecular machinery to its intended destination - the gene you want to regulate. A gRNA is a short RNA sequence designed to be complementary to a specific region of the gene's promoter or coding sequence. This complementarity is essential for ensuring that the CRISPR system homes in on the desired target with remarkable accuracy.

For example, let's say we want to reduce the expression of a gene associated with a certain disease. The gRNA designed for this purpose is crafted to match the genetic sequence of that specific gene. Once inside the cell, the gRNA forms a complementary base pair with the gene's DNA, creating a complex that serves as a binding site for the CRISPR-associated protein, often dCas9 (dead Cas9). This binding is where the magic of gene regulation begins.

Dead but Not Useless: The Role of dCas9

The name "dead" Cas9 might sound somewhat paradoxical, but this protein's role is vital in the CRISPRi process. Unlike its active counterpart, which can cut DNA and initiate repair processes, dCas9 is unable to perform these actions. Instead, it acts as a molecular scaffold, ensuring that the CRISPRi machinery is precisely positioned at the target site.

dCas9's importance lies in its ability to bind to the gRNA and then secure itself onto the gene's DNA at the chosen location. This placement, guided by the gRNA, provides a stable platform for the subsequent regulatory steps. The gRNA-dCas9 complex acts as the surveillance system, with dCas9 serving as the guardian at the gate, preventing unwanted gene expression.

An apt analogy is to consider dCas9 as a lock on the gene's DNA sequence and the gRNA as the key. When they come together, they form a strong, specific bond, essentially locking the gene's expression at a certain level. It's the genetic equivalent of putting the gene "under house arrest" - it can't get away, but it can still function at a reduced pace.

Fine-Tuning with Promoter and Coding Sequence Targeting

CRISPRi is versatile, offering researchers the ability to target both the promoter and coding sequences of a gene, allowing for fine-tuned control. The promoter is the region where RNA polymerase binds to initiate transcription, while the coding sequence contains the information for producing the protein. By focusing on the promoter, researchers can reduce the rate of transcription initiation, effectively silencing the gene at the beginning of the process. In contrast, targeting the coding sequence can lead to incomplete or faulty protein production.

For example, if one is interested in controlling a gene that produces a protein harmful to the cell or organism, CRISPRi can be employed to selectively reduce the gene's expression by targeting the promoter. This results in less RNA being produced, leading to a decrease in the final protein product. Alternatively, researchers might choose to target the coding sequence if their aim is to disrupt the protein's functionality.

This can have profound implications in diseases where the protein in question is malfunctioning or toxic.

To exemplify, consider a gene associated with the development of certain cancers. By designing a gRNA complementary to the promoter region of this gene, researchers can significantly reduce its transcription rate, consequently hampering the production of cancer-promoting proteins. This is a highly precise and controlled approach to gene regulation with the potential to be developed into cancer therapies.

Combining Multiple gRNAs: Multiplexing for Enhanced Control

While individual gRNAs are incredibly potent, the ability to use multiple gRNAs simultaneously, a technique known as multiplexing, provides an additional layer of control. Multiplexing allows researchers to orchestrate the regulation of multiple genes, either individually or collectively. This approach can be likened to conducting a symphony, where each gene's expression is an instrument, and the researcher is the maestro.

Let's illustrate this with a scenario involving a complex biological pathway. Consider a signaling cascade where multiple genes need to be regulated to achieve a desired cellular response. By designing a set of gRNAs, each targeting a different gene within this pathway, researchers can delicately control the entire process. It's like pulling the strings of a marionette, ensuring that all the genes in the pathway dance to a precise tune.

For instance, in the context of neurodegenerative diseases, multiple genes might contribute to the development of the condition. Researchers can employ multiplexing to reduce the expression of these genes simultaneously, potentially slowing or even halting the disease progression. It's a testament to the precision and potential therapeutic applications of CRISPRi.

Balancing Act: Achieving Optimal Gene Suppression

While CRISPRi is remarkable in its precision, there's an art to finding the optimal level of gene suppression. Overly aggressive regulation can lead to unintended consequences, while insufficient suppression might not yield the desired effect. The key is to strike a balance, akin to adjusting the volume on a stereo to achieve the perfect harmony.

In practice, researchers must fine-tune the strength of the CRISPRi system. This can be achieved by adjusting the concentration of dCas9 and the gRNA. Higher concentrations result in more robust gene suppression, while lower concentrations lead to milder regulation. It's a bit like adjusting the pressure on a gas pedal - too much, and you accelerate too quickly, too little, and you don't move at the desired pace.

For example, in the context of metabolic disorders, some genes may need to be suppressed more than others to restore a balance in the system. CRISPRi allows for this level of control. By carefully titrating the CRISPRi components, researchers can dial up or down the expression of specific genes to achieve metabolic equilibrium, potentially offering hope to those with conditions like diabetes or obesity.

Final Thoughts

The precision of CRISPRi is akin to a surgeon's scalpel in the world of gene regulation. It enables researchers to selectively target and modulate individual genes with remarkable accuracy. The marriage of guide RNAs and the dCas9 protein serves as the foundation for this precision, ensuring that the gene of interest is influenced while leaving the rest of the genomic landscape untouched. This level of control, coupled with the capacity for multiplexing and fine-tuning, opens up exciting possibilities for the study and manipulation of gene expression in various fields, from medicine to biotechnology. However, it's imperative to remember that with great power comes great responsibility, and the careful calibration of CRISPRi's precision is essential to achieve desired outcomes without unintended consequences.

2.3 Inhibition of Transcription with CRISPRi

In the vast concerto of life's biological processes, the regulation of gene expression stands as one of the central conductors, directing the orchestra of life. The ability to orchestrate the genes, to silence or enhance their activities, is at the heart of understanding and manipulating life's intricate dance. Within this context, CRISPR interference (CRISPRi) emerges as a powerful and versatile conductor's baton, allowing researchers to wield precise control over the transcription of genes. This section elucidates the underlying mechanisms of how CRISPRi inhibits transcription and showcases the impact of this technique in the realm of gene regulation.

A Molecular Duel: dCas9 vs. Transcription Machinery

At the core of CRISPRi's transcriptional regulation lies a captivating molecular duel. The protagonist is dCas9, a mutated form of the well-known Cas9 protein. Unlike its infamous cousin, which wields the power to cleave DNA, dCas9 brandishes a deactivated blade, rendering it harmless in terms of DNA cleavage. However, it retains the crucial ability to bind to specific DNA sequences with the guidance of a single guide RNA (sgRNA). This binding occurs through Watson-Crick base pairing, just like a key fitting into a lock. Once dCas9 locks onto a target gene's promoter region, it effectively blocks the initiation of transcription.

Here, it's essential to appreciate the marvel of molecular recognition. Each gene's promoter region acts as a unique "lock," and dCas9 with its sgRNA acts as the corresponding "key." This precise recognition is akin to a maestro reading a musical score with exceptional acumen, playing only the intended notes, silencing the rest.

Fine-Tuning Gene Expression: The Role of the sgRNA

While dCas9 is a versatile and steadfast performer, it is the single guide RNA (sgRNA) that imparts specificity to the system. The sgRNA serves as the musical notation, directing dCas9 to its target gene promoter region. This sgRNA is engineered to be complementary to the DNA sequence of the gene one aims to regulate. Think of sgRNA as a conductor's baton, directing the orchestra of transcription machinery to play or pause a particular genetic melody.

The exquisite specificity of sgRNA is worth emphasizing. In a vast genome, each gene's promoter sequence is distinct, like a unique stanza in a grand symphony. The sgRNA ensures that dCas9 selectively silences the chosen gene, avoiding off-target effects. It's akin to a conductor guiding the orchestra to play only the notes on the score, maintaining the purity of the musical performance.

Silencing the Orchestra: CRISPRi Blocks Transcription Initiation

With dCas9 and sgRNA in harmony, CRISPRi can effectively silence a gene's transcription. The process begins at the gene's promoter region, which acts as the transcription start site. In the absence of dCas9, the promoter region serves as a gateway through which the RNA polymerase—the conductor of transcription—gains access to the DNA.

When dCas9, guided by sgRNA, latches onto the promoter region, it forms a steric hindrance, akin to a locked gate barring the entry of RNA polymerase. Without the conductor, the orchestra cannot play, and thus, the gene's transcription grinds to a halt.

This mechanism is akin to a maestro strategically positioning a mute on a trumpet, muffling its sound. In the context of CRISPRi, dCas9 strategically positions itself on the promoter, muting the gene's expression without altering the underlying genetic composition. The beauty of this technique lies in its reversibility and precision—much like removing the mute from the trumpet, transcription can be resumed when needed by merely retracting dCas9.

Real-World Applications of CRISPRi in Transcriptional Inhibition

CRISPRi's capacity to inhibit transcription has opened doors to an array of research applications. Scientists across various fields have harnessed this technology to better understand, manipulate, and potentially treat diseases.

One notable application is in the study of essential genes. These are genes that an organism cannot survive without. Conventional gene knockout techniques that permanently deactivate a gene often lead to cell death, making the study of essential genes challenging. With CRISPRi, the temporary inhibition of transcription allows researchers to investigate these genes' functions without causing irreversible damage, expanding our knowledge of vital biological processes.

Moreover, CRISPRi's precision is invaluable in dissecting complex regulatory networks. It enables researchers to selectively silence multiple genes within a pathway, observing the effects on overall gene expression and phenotype. This dissect-and-observe approach is akin to isolating sections of an orchestra to understand their individual contributions to the music while keeping the rest of the performance intact.

In the field of medicine, CRISPRi shows promise in the study of genetic diseases. By temporarily silencing disease-associated genes, researchers can explore potential therapeutic strategies and gain insights into the molecular basis of these disorders. This approach is like selectively muting a discordant note in an otherwise harmonious composition, striving to restore the symphony of health.

Beyond Genes: Epigenetic Regulation with CRISPRi

While CRISPRi's primary role is in gene transcription, its influence extends beyond the genetic code. It has been adapted to target and modify epigenetic marks, which are chemical modifications on the DNA and histones that influence gene expression. By utilizing dCas9 fused with epigenetic modifiers, researchers can precisely alter these marks, turning genes on or off at an epigenetic level. This capability is analogous to a conductor not only influencing the musicians' notes but also altering the acoustics of the concert hall to shape the overall performance.

In short, CRISPRi's capacity to inhibit transcription is a remarkable example of molecular precision in the field of gene regulation. It operates with the finesse of a skilled conductor, orchestrating gene expression while maintaining specificity and reversibility. This technology's real-world applications are expanding our understanding of biology, offering insights into diseases, and providing potential avenues for therapeutic interventions. Moreover, its adaptability for epigenetic regulation adds a new layer of complexity to the symphony of life's gene expression. As we continue to refine and explore the potential of CRISPRi, its role in shaping the future of biology and medicine remains a captivating melody waiting to be composed.

Chapter 3: Applications of CRISPRi in Basic Research

3.1 Studying Gene Function

Gene function represents the cornerstone of molecular biology, and understanding how genes work is fundamental for unravelling the mysteries of life. The advent of CRISPRi has revolutionized the way we investigate gene function, providing scientists with powerful tools to dissect the intricacies of gene activity in a targeted and precise manner. In this section, we will explore how CRISPRi is being harnessed to shed light on the functionality of genes and the broader implications this has for basic research.

The Traditional Approach: Knockouts and RNA Interference (RNAi)

Before delving into the specifics of CRISPRi, it's essential to appreciate the traditional methods used for studying gene function. One such method is gene knockout, which entails the complete removal or inactivation of a gene. Knockout studies have provided invaluable insights into the essentiality of specific genes. In contrast, RNA interference (RNAi) involves the introduction of short RNA molecules to inhibit the translation or stability of target gene transcripts. While these methods have their merits, they do come with certain limitations.

One of the drawbacks of gene knockout is that it often results in permanent gene loss. This limitation makes it challenging to study genes that might have essential functions during specific developmental stages or in particular tissues, as complete gene loss can be lethal. On the other hand, RNAi, while less permanent, can have off-target effects, potentially interfering with genes other than the intended target.

CRISPRi: A Precision Scalpel for Gene Function Studies

CRISPR interference (CRISPRi) stands out as a game-changer in the field of gene function studies. At its core, CRISPRi employs a modified version of the CRISPR-Cas9 system. Instead of completely removing a gene or introducing foreign RNA molecules, CRISPRi allows researchers to subtly modulate gene expression levels with exceptional precision.

Example 1: Identifying Gene Function in Development

Consider a scenario where scientists aim to understand the role of a specific gene in the development of an organism. With CRISPRi, they can finely tune the expression of this gene rather than knocking it out entirely. By doing so, they may discover that the gene is crucial at certain stages of development but dispensable in others. This level of control enables a more nuanced understanding of gene function within the intricate dance of development.

Example 2: Investigating Gene Interaction Networks

In molecular biology, genes seldom act in isolation; they often function as part of complex networks. CRISPRi's ability to modulate gene expression in a reversible manner is a boon for studying these interactions. Researchers can dial genes up or down and observe the ripple effects on the entire network. This dynamic approach has led to the discovery of novel pathways and interactions that might have been missed through traditional knockout methods.

The Power of Temporal Control

One of the distinguishing features of CRISPRi is its temporal control. Unlike gene knockout, which is typically a one-time event, CRISPRi allows researchers to turn gene expression up or down at different time points. This temporal control is invaluable for studying genes involved in processes that change over time.

Example 3: Unravelling the Secrets of Circadian Rhythms

Circadian rhythms, the biological clocks that regulate our daily lives, are driven by a complex web of genes. CRISPRi has been instrumental in understanding these rhythms by enabling the manipulation of gene expression in a time-dependent manner. Researchers can simulate the effects of gene expression changes at various times of day, uncovering the intricate regulatory mechanisms that govern our sleep-wake cycles and physiological processes.

Precision in Tissue-Specific Studies

Another remarkable facet of CRISPRi is its ability to target specific tissues or organs within an organism. This feature is particularly advantageous when studying genes with multifaceted roles throughout the body.

Example 4: Dissecting Heart Development

Imagine the study of a gene vital for heart development. Rather than affecting the entire organism, CRISPRi allows researchers to pinpoint the heart tissue. This targeted approach can help in ascertaining the gene's specific role in cardiac development without causing potentially lethal consequences in other tissues.

A Potential Solution to Off-Target Effects

Off-target effects are a notorious concern in genetic manipulation techniques. CRISPRi, while not entirely exempt from this issue, has shown promise in minimizing unintended consequences.

Example 5: Minimizing Off-Target Effects in Cancer Research

In cancer research, the precision of CRISPRi is especially beneficial. By carefully modulating the expression of genes associated with tumour growth, researchers can avoid affecting non-cancerous cells. This not only leads to a more accurate representation of the gene's role in cancer but also reduces the risk of unwanted side effects in potential therapeutic applications.

Transcending Species Barriers

The versatility of CRISPRi extends to its applicability across diverse species. This adaptability enables comparative studies across different organisms, providing insights into gene conservation and divergence.

Example 6: Comparative Genomics of Eye Development

Researchers interested in the development of the eye have used CRISPRi to study the genes involved in this process in various species, from fruit flies to mice. By comparing the effects of gene modulation in different organisms, they can uncover the fundamental genetic principles governing eye development and the unique adaptations that have occurred in each species.

Interrogating Non-Coding Elements

Gene function is not limited to protein-coding genes. Non-coding elements, such as long non-coding RNAs and enhancers, also play crucial roles in gene regulation. CRISPRi's precision extends to these elements.

Example 7: Investigating Non-Coding RNAs in Neurological Disorders

In neurobiology, researchers have used CRISPRi to explore the function of non-coding RNAs implicated in neurological disorders. By selectively inhibiting these RNAs, they can decipher their roles in brain development and disease, potentially uncovering new therapeutic targets.

Enhancing Reproducibility and Rigor

In an era where reproducibility is a paramount concern, the fine-tuned control that CRISPRi offers can significantly enhance the rigor of scientific experiments. Experiments can be precisely replicated, and results can be more confidently attributed to specific genetic modifications.

Final Thoughts

CRISPRi has truly transformed the way we study gene function. Its precision, temporal control, and tissue specificity offer a level of insight and versatility that was previously unimaginable. By providing a nuanced understanding of how genes operate in various contexts, CRISPRi is propelling the field of basic research to new heights, with profound implications for our understanding of biology and the potential for innovative breakthroughs in various scientific disciplines.

3.2 Functional Genomics with CRISPRi

This section explores the remarkable power of CRISPRi in deciphering the complex genetic code, illuminating previously hidden facets of gene function, and paving the way for breakthroughs in our understanding of biology.

Unlocking the Genome's Mysteries

Genomics, as a scientific attempt, has consistently sought to unlock the secrets of the genetic drapery that underpins life. While the DNA sequence was meticulously decoded in the early 2000s through colossal efforts like the Human Genome Project, our ability to comprehend the functions of individual genes lagged behind. This is where CRISPRi takes centre stage.

The Gene Regulation Puzzle

In the vast landscape of genomics, gene regulation emerges as a critical piece of the puzzle. Deciphering how genes are turned on or off, and to what extent, is integral to understanding the orchestration of life processes. Traditional genetic tools provided a snapshot, allowing researchers to identify genes implicated in specific traits or diseases. However, CRISPRi offers the unique ability to fine-tune gene expression, enabling a more dynamic understanding of gene function.

CRISPRi's Precision Instrumentation

At the core of CRISPRi's functional genomics capability lies its exquisite precision. This technology hinges on a minimalist yet powerful setup: the Cas9 nuclease, the universal workhorse of CRISPR-based systems, is defanged to become a programmable gene silencer. By mutating Cas9 into a catalytically inactive form (dCas9), it can be guided with RNA molecules to virtually any genomic location of interest. This enables the selective repression of target genes with pinpoint accuracy.

Functional Genomics in Action

To appreciate the transformative potential of CRISPRi in functional genomics, consider its application in a common research scenario: the identification of essential genes in a model organism like the bacterium Escherichia coli (E. coli).

In traditional genetic studies, the approach involved knocking out genes to observe the effects. However, the total deletion of essential genes would result in a non-viable organism, rendering the analysis impossible. This is where CRISPRi shines. By gently turning down the expression of essential genes, researchers can gauge the consequences without incapacitating the organism. It's akin to dimming a spotlight rather than extinguishing it altogether.

This ability to modulate gene expression provides an invaluable tool for systematically exploring the functions of genes across the genome. Large-scale CRISPRi screens can be employed to assess the impact of gene repression on cellular behaviours. For instance, researchers can uncover genes responsible for resistance to antibiotics, genes involved in stress responses, or genes that influence growth rates.

Beyond Bacteria: Functional Genomics in Eukaryotes

While the E. coli example illustrates CRISPRi's efficacy in prokaryotes, its transformative potential extends to eukaryotes, including humans. In eukaryotic systems, CRISPRi allows researchers to explore the functions of genes in a manner that was previously challenging.

For instance, when studying human genes, the approach isn't as straightforward as it is in bacteria. Knocking out essential genes in human cells often leads to cell death, making functional analysis complicated. With CRISPRi, researchers can precisely regulate the activity of these genes, thereby avoiding the catastrophic consequences of their complete removal. This has paved the way for a deeper understanding of human gene function.

CRISPRi Screens: A Genomic Treasure Hunt

One of the most exciting applications of CRISPRi in functional genomics is the implementation of genome-wide screens. These screens, which involve systematically targeting every gene in the genome to assess its function, have transformed our ability to unravel the genetic basis of various traits and diseases.

In a typical CRISPRi screen, a library of guide RNAs (gRNAs) is designed to target every gene in the genome individually. Each gRNA is then used to guide dCas9 to a specific gene's promoter region, suppressing its expression. By conducting comprehensive screens in this manner, researchers can uncover genes that are crucial for diverse processes, from cellular growth to the immune response.

For example, a CRISPRi screen in human cells might aim to identify genes that, when suppressed, enhance susceptibility to a viral infection. By systematically silencing each gene and exposing the cells to the virus, researchers can pinpoint genes that play essential roles in the antiviral defence system.

Such screens have had a transformative impact in fields like cancer research, where they've been used to uncover genes that contribute to the uncontrolled growth of cancer cells. In neuroscience, CRISPRi screens have elucidated genes associated with neurodegenerative disorders. These findings open up new avenues for drug development and therapeutic strategies.

CRISPRi and Functional Genomics: A Symbiotic Relationship

In essence, CRISPRi and functional genomics have entered into a symbiotic relationship. The precision of CRISPRi empowers functional genomics to dissect the intricate machinery of gene regulation, while functional genomics provides CRISPRi with a roadmap to explore the expansive terrain of gene function. Together, they are unravelling the mysteries of the genome with unprecedented detail and accuracy.

The Road Ahead

As the landscape of functional genomics continues to evolve, the CRISPRi toolkit expands. Innovations in gRNA design, delivery methods, and improved understanding of off-target effects enhance the technology's efficacy. The future promises more comprehensive screens, deeper insights into gene function, and breakthroughs that will impact fields ranging from medicine to agriculture.

In the next chapter, we delve into the tools and techniques that underpin the success of CRISPRi, offering a closer look at the design and delivery of these precision instruments.

3.3 Dissecting Complex Biological Pathways

In the world of molecular biology, understanding the intricacies of complex biological pathways is akin to deciphering an ancient cryptic script. Each gene, each protein, each interaction, contributes to the orchestration of life itself. This chapter explores how CRISPRi, with its precision and versatility, has emerged as a formidable tool to tackle the challenge of dissecting these complex biological pathways.

Unraveling the Complexity

The understanding of biology, in all its resplendent complexity, has been a central pursuit of scientists for centuries. It is akin to unraveling the pages of a grand epic – a tale written in the language of DNA and proteins. To navigate through this intricate narrative, researchers need tools that offer both precision and versatility. This is where CRISPRi comes into play, as it allows scientists to selectively dampen the expression of specific genes, shedding light on their roles within these intricate storylines.

Gene Function and Regulation

Biological pathways often resemble intricate machinery where genes serve as individual cogs and proteins act as gears. To understand the machinery's operation, it is crucial to decipher the role of each component and how they interact. For instance, consider the $p53$ pathway, which is central to regulating the cell cycle and preventing cancer. With CRISPRi, researchers can selectively turn down the expression of specific genes in this pathway, thereby elucidating their influence on the cell's destiny.

The Versatility of CRISPRi

The CRISPR interference system is an invaluable asset in the toolkit of molecular biologists. Unlike its more famous cousin, CRISPR-Cas9, which is a genome-editing powerhouse, CRISPRi focuses on silencing genes without making permanent alterations to the genome. This feature makes it particularly well-suited for the nuanced study of complex pathways.

Fine-Tuning Gene Expression

Intricacy in biology often arises from the precise regulation of genes. Some genes need to be expressed at specific times or in specific cell types, while others must be held in check. This balance is akin to an orchestra conductor ensuring that each instrument plays its part at the right moment. CRISPRi serves as the maestro's wand, allowing scientists to fine-tune gene expression by selectively silencing or repressing specific genes, providing a newfound level of control in studying complex pathways.

Specificity in Action

One of the remarkable aspects of CRISPRi is its specificity. Just as a locksmith carefully selects the right key for a particular lock, researchers design guide RNAs that precisely target the genes of interest. This precision enables the isolation of individual components within a biological pathway. For example, in the case of the intricate Notch signalling pathway, which plays pivotal roles in development and disease, CRISPRi can be used to silence specific genes, thus allowing the study of their contributions to the broader network.

Functional Genomics with CRISPRi

In the age of genomics, understanding the function of every gene within an organism's DNA is an ongoing endeavor.

However, it's not just about identifying genes; it's about comprehending their roles in the grand scheme of life. CRISPRi, with its ability to silence genes selectively, empowers researchers to perform what can be considered 'genetic surgery,' where they carefully modulate gene expression to see how it impacts the whole organism or specific biological pathways.

Case Study: Notch Signalling Pathway

The Notch pathway exemplifies a biological network of remarkable complexity. It regulates critical processes such as cell fate determination, proliferation, and differentiation. Dysregulation of this pathway is linked to several diseases, including cancer. By employing CRISPRi, scientists have been able to isolate and investigate key components within the Notch pathway.

For instance, a specific study focused on silencing the Notch1 gene using CRISPRi in human lung adenocarcinoma cells. By reducing the expression of Notch1, researchers observed a decrease in cell proliferation and an increase in apoptosis. This experiment provided a clear link between Notch1 and the growth of cancerous cells within the lung, underscoring the pathway's significance in the context of cancer biology.

Genetic Crossroads

Biological pathways, like intricate mazes, often present multiple crossroads and junctions, where one gene can influence various outcomes. Here, CRISPRi emerges as the guiding compass. It allows scientists to navigate these crossroads by specifically inhibiting individual genes and observing the consequences, helping to map out the pathways and interactions in unprecedented detail.

CRISPRi and Cellular Memory

In cellular biology, 'memory' refers to the phenomenon where a cell retains information about its previous states and experiences. This memory is often encoded in the epigenetic modifications of genes, influencing how they respond to signals and cues. CRISPRi can be used to dissect the machinery of cellular memory, revealing the genes responsible for retaining and transmitting information within complex biological pathways.

Infectious Disease Research

CRISPRi has also found application in untying the complex relationship between pathogens and their hosts. For instance, in the context of viral infections, understanding how viruses manipulate host cell machinery is essential. CRISPRi allows researchers to selectively silence host genes and observe how this impacts the virus's ability to replicate and infect. This approach has provided insights into the molecular intricacies of host-pathogen interactions.

Chromatin Dynamics

Within the nucleus of a cell, DNA is not a static entity but a dynamic player in gene regulation. Chromatin, the complex of DNA and proteins, plays a pivotal role in determining which genes are active and which remain silent. CRISPRi can be employed to investigate the complex dynamics of chromatin, aiding in the understanding of how genes are turned on and off in response to various signals.

Final Thoughts

CRISPRi, with its precision and versatility, has revolutionized the way researchers investigate complex biological pathways. By selectively silencing genes within intricate networks, scientists can uncover the roles and interactions of individual components, shedding light on the grand narrative of life encoded in DNA and proteins. This technology is not just a tool but a beacon that guides researchers through the convoluted passages of biology, helping us decipher the language of life itself.

Chapter 4: Tools and Techniques for CRISPRi

4.1 Designing Effective gRNAs

The remarkable efficacy of CRISPR interference (CRISPRi) hinges on the precision and efficiency of its guide RNAs (gRNAs). These molecular compasses play a pivotal role in navigating the CRISPR-Cas9 system to its target, thereby determining the success or failure of gene regulation. This section delves into the intricacies of designing effective gRNAs, and their pivotal role in CRISPRi applications.

The Crucial Role of gRNAs

gRNAs, or guide RNAs, are the sentinels of the CRISPRi system. These are single-stranded RNA molecules with a simple but indispensable task: guiding the Cas9 protein to the desired genomic location. The Cas9 protein, often referred to as a 'molecular scissor,' can effectively cut DNA, but in CRISPRi, it remains inactive and acts solely as a search engine, thanks to the guidance provided by the gRNA.

The importance of designing effective gRNAs cannot be overstated. These molecules are the linchpin of CRISPRi gene regulation and must be tailored with precision. A poorly designed gRNA can lead to off-target effects, inefficient gene silencing, and a slew of technical challenges. Conversely, a well-designed gRNA can make CRISPRi a powerful tool for specific and controlled gene regulation.

Principles of gRNA Design

The design of gRNAs revolves around several key principles, each of which contributes to the success and specificity of the CRISPRi system.

Target Selection: The starting point for designing an effective gRNA is the careful selection of the target gene. The target gene should be chosen based on its relevance to the research question or application. Researchers must consider the gene's function, importance, and the specific outcomes they aim to achieve through gene regulation. A clear understanding of the genetic context is essential to guide the choice of the target site.

Sequence Specificity: The gRNA must be tailored to be highly specific to the target gene. This specificity is determined by the sequence of the gRNA, which should be meticulously designed to match the target DNA sequence. Advances in computational tools have made it feasible to identify target sequences with minimal or no homology to other genomic regions. These tools analyse the entire genome to predict and avoid off-target binding.

Protospacer Adjacent Motif (PAM): A CRISPRi gRNA must also include a Protospacer Adjacent Motif (PAM) sequence. The PAM is a short, invariant sequence recognized by the Cas9 protein and is essential for initiating the binding process. The choice of PAM sequence depends on the specific Cas9 variant being used, and it's important to select a PAM that is both compatible with the chosen Cas9 and strategically positioned to enable effective gene regulation.

Secondary Structure and Folding: The gRNA's secondary structure and folding are critical considerations. A gRNA should not contain extensive secondary structures or sequences prone to self-binding, as this can impede its function. Computational tools can predict potential secondary structures, aiding in the selection of gRNAs with minimal interference from internal base pairing.

gRNA Length: The length of a gRNA is another critical factor in design. gRNAs are usually 20 nucleotides in length, although shorter gRNAs have been explored. A balance must be struck between gRNA length and specificity. Longer gRNAs may increase specificity but could also be less efficient due to steric hindrance, while shorter gRNAs might result in off-target effects.

Off-Target Analysis: A robust gRNA design process involves an in-depth analysis of potential off-target sites in the genome. The goal is to minimize off-target effects, which can lead to unintended gene regulation. Various algorithms and tools are available to predict and assess potential off-target binding sites for a given gRNA sequence.

Empirical Validation: While computational tools can greatly aid gRNA design, empirical validation remains a crucial step. Experimental testing of designed gRNAs in the target organism is necessary to confirm their efficacy and specificity. This empirical validation is a safeguard against unforeseen off-target effects that might not be predicted by computational tools alone.

Improving gRNA Design: Challenges and Innovations

The field of gRNA design is ever-evolving, and researchers continue to refine and innovate their approaches to create more effective gRNAs. Here are some of the challenges and recent advancements in gRNA design:

Off-Target Effects: Despite significant progress in reducing off-target effects, they still pose a challenge in CRISPRi applications. Recent innovations involve the development of high-fidelity Cas9 variants and modified gRNAs to enhance specificity. These innovations hold promise for minimizing off-target effects.

Multiplexing: Designing gRNAs for multiplexed gene regulation, where multiple genes are targeted simultaneously, requires careful consideration of each gRNA's specificity and efficiency. Advances in multiplex gRNA design have paved the way for more complex and powerful gene regulation strategies.

Delivery Methods: The choice of delivery method for gRNAs can impact their design. Different delivery methods, such as viral vectors or nanoparticles, require specific gRNA designs to ensure efficient and precise gene regulation.

In Vivo Applications: In vivo applications of CRISPRi demand gRNAs that work effectively within the context of a living organism. Design considerations must include factors like gRNA stability, tissue-specific delivery, and immune responses.

Ethical Considerations: With the potential for misuse of CRISPR technology, ethical considerations in gRNA design include strategies to prevent or limit the use of CRISPRi for unethical purposes.

Final Thoughts

Designing effective gRNAs is a pivotal step in the CRISPRi gene regulation process. It involves a delicate interplay of molecular biology, computational analysis, and empirical validation. Success in CRISPRi applications hinges on the ability to design gRNAs that are not only specific to the target gene but also free from off-target effects. Ongoing research and innovations in gRNA design promise to further enhance the precision and power of CRISPRi as a gene regulation tool, opening new avenues in biotechnology, medicine, and fundamental research.

4.2 Delivery Methods for CRISPRi

The successful implementation of CRISPR interference (CRISPRi) hinges not only on the precision of its molecular components but also on the effectiveness of delivery methods. In this section, we explore the diverse array of delivery techniques employed to introduce the necessary CRISPRi components, dCas9 and guide RNA (gRNA), into target cells and organisms. From viral vectors to chemical transfection, each method presents its own set of advantages and challenges, ultimately shaping the adaptability of CRISPRi for various applications.

Viral Vectors: A Trusted Workhorse

Viral vectors have emerged as one of the most trusted and widely employed delivery systems for CRISPRi. They offer several advantages, including high transfection efficiency and the ability to target specific cell types. Notably, lentiviruses and adeno-associated viruses (AAVs) have found extensive use in delivering CRISPRi components.

Lentiviral Vectors

Lentiviral vectors are derived from the human immunodeficiency virus (HIV) but are modified to be replication-incompetent and safe for research and therapeutic purposes. Their integration into the host genome allows stable and long-term expression of dCas9 and gRNA. This makes them an attractive option for sustained gene regulation.

In a seminal study published in *Nature Methods* by Gilbert et al. in 2014, lentiviral vectors were used to deliver dCas9 and gRNA, leading to robust gene silencing in human cells. The study demonstrated the feasibility of CRISPRi in diverse cell types and laid the foundation for subsequent investigations.

Adeno-Associated Viruses (AAVs)

AAVs are another viral vector of choice due to their low immunogenicity and the ability to infect both dividing and non-dividing cells. These vectors have gained popularity for in vivo applications, particularly in animal models and potential gene therapies.

One noteworthy example is a study conducted by Kiani et al. and published in *Cell* in 2015. The researchers employed AAVs to deliver dCas9 and gRNA for the regulation of the tyrosinase gene in a mouse model, resulting in controlled pigmentation. This demonstrated the potential of AAVs in achieving targeted gene regulation in living organisms.

While viral vectors offer high transfection efficiency, long-term expression, and cell-type specificity, they are not without limitations. These include the potential for immune responses, limited cargo capacity, and challenges in large-scale production, all of which must be carefully considered in the context of therapeutic applications.

Chemical Transfection: Versatility and Accessibility

Chemical transfection methods provide a versatile and cost-effective alternative to viral vectors for CRISPRi delivery. These methods involve the use of lipids, polymers, or other chemical agents to introduce dCas9 and gRNA into target cells.

Lipofection

Lipofection is a widely used chemical transfection method that utilizes cationic lipids or liposomes to encapsulate and deliver CRISPRi components into cells. Lipofection is relatively simple and can be performed in most laboratories without the need for specialized equipment.

A notable study by Qi et al., published in *Nature Biotechnology* in 2013, demonstrated the use of lipofection to introduce dCas9 and gRNA into human cells, achieving efficient gene repression. The simplicity and accessibility of this method have made it a popular choice for many researchers.

Electroporation

Electroporation, another chemical transfection method, employs brief electrical pulses to create temporary pores in the cell membrane, allowing for the entry of CRISPRi components. It is particularly effective for hard-to-transfect cells and can be scaled up for high-throughput applications.

A study published in *Science* by Gootenberg et al. in 2017 showcased the power of electroporation in delivering CRISPRi into bacterial cells. The researchers used this method to precisely regulate gene expression in a diverse range of bacterial species, expanding the potential applications of CRISPRi in microbiology.

Chemical transfection methods offer versatility and are less likely to trigger immune responses than viral vectors. However, they may have lower transfection efficiency and be less suitable for in vivo applications due to potential toxicity concerns.

Nanoparticles and Nanotechnology

Nanoparticles and nanotechnology-based delivery systems have gained attention for their potential to enhance the precision and efficiency of CRISPRi delivery. These systems take advantage of nanoscale carriers to protect and transport CRISPRi components to their intended destinations.

Gold Nanoparticles

Gold nanoparticles have been explored as carriers for CRISPRi due to their stability and ease of functionalization. By attaching dCas9 and gRNA to gold nanoparticles, researchers can enhance the cellular uptake and gene regulation capabilities of CRISPRi.

In a study published in *ACS Nano* in 2016, researchers developed a gold nanoparticle-based system for CRISPRi delivery into neurons. The system exhibited promising results in achieving gene regulation in a neuronal context, which has implications for neurological research and potential therapeutic interventions.

Lipid Nanoparticles

Lipid nanoparticles, or lipid-based nanocarriers, offer a lipid-based approach similar to lipofection but on a nanoscale. These nanoparticles can encapsulate and protect CRISPRi components, facilitating their cellular delivery.

A study by Finn et al., published in *Molecular Therapy* in 2018, utilized lipid nanoparticles to deliver CRISPRi components for the treatment of a mouse model of Duchenne muscular dystrophy. The researchers demonstrated the potential of lipid nanoparticles in achieving therapeutic gene regulation in a complex in vivo setting.

While nanoparticle-based delivery systems show promise in improving the precision and efficiency of CRISPRi, further research is needed to optimize their design and ensure their safety for therapeutic applications.

Microinjection: A Precise but Labor-Intensive Approach

Microinjection is a highly precise but labour-intensive method of CRISPRi delivery. This technique involves the direct injection of CRISPRi components into the target cells or embryos, making it suitable for a variety of applications.

Intracytoplasmic Sperm Injection (ICSI)

In the field of reproductive biology, intracytoplasmic sperm injection (ICSI) is a microinjection method that has been adapted for CRISPRi. It allows for the introduction of CRISPRi components into fertilized eggs, offering the potential to edit genes in the resulting offspring.

A study by Kanca et al., published in *Cell* in 2019, used ICSI to introduce CRISPRi components into Drosophila embryos. This approach enabled the precise regulation of gene expression in developing organisms, demonstrating the versatility of microinjection in model organisms.

Single-Cell Microinjection

Single-cell microinjection is a powerful tool for studying gene regulation at the single-cell level. It involves the injection of CRISPRi components into individual cells, enabling researchers to dissect gene expression patterns and cellular behaviours with high precision.

A study published in *Nature* by Datlinger et al. in 2017 showcased the use of single-cell microinjection to investigate the regulatory landscape of individual cells in human tissues. This approach provided insights into the heterogeneity of gene expression and the potential for therapeutic interventions at the single-cell level.

Microinjection is advantageous for its precision, making it suitable for applications requiring high spatial and temporal control. However, it is a labour-intensive and technically demanding method, limiting its scalability for certain research and therapeutic purposes.

In Vivo and In Situ Delivery

In vivo and in situ CRISPRi delivery methods are designed for applications within living organisms or tissues, offering the potential for gene regulation at the organismal level.

Hydrodynamic Injection

Hydrodynamic injection is a technique used for the in vivo delivery of CRISPRi components, primarily in animal models. This method involves the rapid injection of a large volume of solution containing CRISPRi components into the circulatory system of the animal.

A study by Yin et al., published in *Nature Communications* in 2016, employed hydrodynamic injection to deliver CRISPRi components for gene regulation in the livers of mice. The approach allowed for the modulation of gene expression in a tissue-specific manner, with potential implications for metabolic research and gene therapy.

Topical Application

In situ gene regulation can be achieved through topical application of CRISPRi components. This approach is particularly relevant for skin-related applications and localized gene regulation.

A study published in *Science Translational Medicine* by Wu et al. in 2017 demonstrated the use of topical application of CRISPRi components to regulate gene expression in the skin of mice. The researchers achieved targeted gene silencing with the potential for dermatological treatments and cosmetic applications.

In vivo and in situ delivery methods are essential for applications involving whole organisms and localized gene regulation. However, these methods may require optimization for specific tissues and may present challenges in ensuring precise targeting and controlled release.

Future Perspectives in CRISPRi Delivery

As the field of CRISPRi continues to evolve, so too do the methods of delivery. Researchers are actively exploring new strategies to enhance the precision, safety, and efficiency of introducing CRISPRi components into target cells and organisms.

Advancements in nanotechnology hold promise for the development of more sophisticated nanoparticles and nanocarriers that can navigate biological barriers with greater ease. Additionally, the design of synthetic delivery systems, such as cell-penetrating peptides and protein transduction domains, is a growing area of interest.

Furthermore, the combination of delivery methods, such as the use of viral vectors for initial delivery and subsequent fine-tuning with chemical transfection, may offer a synergistic approach to optimize CRISPRi regulation.

Therefore, the success of CRISPRi is inherently linked to the effectiveness of its delivery methods. The choice of delivery system depends on the specific needs of the research or therapeutic application, and each method presents its own set of advantages and challenges. As the CRISPRi field advances, innovative delivery strategies will continue to broaden the scope of gene regulation, making it increasingly accessible and adaptable for a wide range of scientific events and potential clinical applications.

4.3 Off-Target Effects and Minimizing Non-Specific Interference

In our exploration of CRISPR interference (CRISPRi) as a gene regulation tool, it's crucial to address one of the most pressing challenges associated with this technology: off-target effects. While CRISPRi offers remarkable precision in manipulating gene expression, unintended disruptions at non-target sites can undermine its reliability and safety. To harness the full potential of CRISPRi, researchers have been diligently working to minimize non-specific interference.

Understanding Off-Target Effects

Off-target effects are the unintended consequences of CRISPRi, where the dCas9 protein and guide RNA (gRNA) may inadvertently bind to sequences resembling the intended target, leading to unintended gene regulation. These off-target effects are undesirable as they can introduce unpredictable changes in the cellular environment and potentially compromise experimental results or therapeutic applications.

Numerous studies have highlighted the existence of off-target effects in CRISPR-based technologies. For instance, a study published in the journal *Science* in 2018 illustrated the off-target activity of CRISPR-Cas9, a close relative of CRISPRi. The researchers revealed that off-target effects, if left unaddressed, could lead to unintended genetic alterations that might have adverse consequences in therapeutic contexts. Hence, to harness CRISPRi effectively, mitigating off-target effects is of paramount importance.

Improving gRNA Design

To mitigate off-target effects, one fundamental approach is to enhance the design of gRNAs. gRNAs serve as molecular GPS, guiding dCas9 to its target gene. Improved gRNA design is the first line of defence in ensuring specificity.

Recent advancements in bioinformatics have made it possible to predict potential off-target sites with higher accuracy. For example, the development of algorithms like MIT CRISPR and CRISPOR empowers researchers to identify sequences with similarities to the intended target. These tools enable scientists to select gRNAs with fewer off-target predictions, thereby reducing the likelihood of unintended gene regulation.

Moreover, the use of shorter gRNAs can enhance specificity. Research conducted by Mali et al. in 2013 demonstrated that truncated gRNAs led to a significant reduction in off-target effects. By shortening the gRNA length, researchers can limit the number of potential off-target sites, ensuring that dCas9 is more likely to bind to the intended target gene.

Enhanced dCas9 Variants

Another strategy to minimize off-target effects is the development of enhanced dCas9 variants. While traditional dCas9 proteins may exhibit some off-target binding, modified variants have been engineered to increase specificity.

For example, the "high-fidelity" or "nickase" Cas9 variant contains point mutations that significantly reduce off-target effects. These mutations enhance the precision of dCas9, making it less likely to bind to sequences with partial similarity to the target site. By using high-fidelity dCas9 variants, researchers can minimize unintended gene regulation while preserving the intended on-target effects.

Epigenetic Modifications for Specificity

Intriguingly, researchers have also explored the use of epigenetic modifications to enhance CRISPRi specificity. Epigenetic marks, such as DNA methylation and histone modifications, play a crucial role in gene regulation by determining whether a gene is active or silenced. Leveraging these marks can help guide dCas9 to its intended target with greater precision.

A study conducted by Thakore et al. in 2015 demonstrated that the fusion of dCas9 with DNA methyltransferases or histone modifiers could direct dCas9 to specific genomic loci based on their epigenetic state. This innovative approach allows researchers to regulate gene expression in a more context-dependent manner, reducing off-target effects significantly.

Utilizing Enhanced Bioinformatics Tools

Advancements in bioinformatics have been instrumental in tackling off-target effects. The development of deep learning algorithms, such as DeepCRISPR, has improved the prediction of off-target sites. These tools leverage large datasets and sophisticated machine learning techniques to provide more accurate assessments of potential off-target binding, thus guiding the selection of gRNAs with higher specificity.

Moreover, off-target analysis has been made more accessible with the advent of genome-wide CRISPRi screens. These screens, such as CROP-Seq, allow researchers to assess the impact of CRISPRi on a broader scale, providing valuable data to evaluate potential off-target effects in a more comprehensive manner.

Experimental Validation of Off-Target Effects

While computational tools have become invaluable in predicting off-target effects, experimental validation remains a critical step in ensuring the specificity of CRISPRi. It is important to validate the effects of CRISPRi at both the intended target and potential off-target sites.

Validation techniques include chromatin immunoprecipitation followed by sequencing (ChIP-seq) and RNA sequencing (RNA-seq). ChIP-seq can determine the binding of dCas9 at specific loci, while RNA-seq can reveal changes in gene expression. Combining computational predictions with experimental data helps researchers confirm the specificity of their CRISPRi system.

Multiplexed CRISPRi for Specificity

To further enhance specificity, researchers have developed multiplexed CRISPRi systems. These systems involve using multiple gRNAs to target the same gene, simultaneously. By targeting the gene from multiple angles, researchers can increase the specificity of gene regulation while reducing the likelihood of off-target effects.

Multiplexed CRISPRi has been particularly valuable in large-scale genetic screens, where targeting multiple genes simultaneously is essential. The approach reduces the impact of off-target effects on the overall results, improving the accuracy of high-throughput experiments.

Final Thoughts

In the journey to harness the potential of CRISPR interference for gene regulation, addressing off-target effects is an ongoing priority. These efforts are integral to the success of CRISPRi, whether in basic research or therapeutic applications.

As the field of CRISPRi continues to evolve, the refinement of gRNA design, the development of enhanced dCas9 variants, the integration of epigenetic modifications, and the utilization of advanced bioinformatics tools collectively contribute to the quest for specificity. Additionally, experimental validation and multiplexed CRISPRi strategies offer promising avenues to enhance the precision of gene regulation.

The path forward in minimizing off-target effects is paved with innovation and collaboration among researchers, driven by the shared goal of realizing the full potential of CRISPRi while ensuring the safety and reliability of this groundbreaking technology. As we move forward, the optimization of CRISPRi for specific and targeted gene regulation promises to unlock a new era in genetics and biomedicine.

Chapter 5: Challenges and Limitations of CRISPRi

5.1 Ethical Considerations in Gene Regulation

As we steer the remarkable terrain of CRISPR interference (CRISPRi) and its potential for gene regulation, we are confronted by ethical considerations that demand thoughtful exploration. The power to precisely manipulate genes brings with it a responsibility to critically assess the implications of such capabilities. In this chapter, we scrutinize the ethical dimensions surrounding gene regulation and delve into the complexities of balancing scientific progress with moral imperatives.

Navigating the Ethical Quandary

The prospect of manipulating genes raises ethical questions that resonate across scientific, societal, and philosophical spectrums. At the forefront is the ethical use of CRISPRi in humans, prompting debates on the fine line between therapeutic intervention and genetic enhancement. The ability to modulate genes associated with intelligence, physical prowess, or even aesthetic traits sparks discussions about the ethical boundaries of tinkering with fundamental aspects of human identity.

Example 1: Germline Editing and the Inheritance Dilemma

One particularly contentious issue revolves around germline editing – the alteration of genes in reproductive cells. While this holds the promise of eliminating hereditary diseases before they manifest, it also introduces the prospect of heritable genetic modifications. The notion of "designer babies," where parents could choose specific traits for their offspring, raises ethical concerns about the commodification of life and the potential exacerbation of social inequalities.

Example 2: The CRISPR Baby Controversy

The ethically charged atmosphere was palpable with the 2018 revelation of the birth of genetically modified twins in China, whose genomes were edited to confer resistance to HIV. This unprecedented move by scientist Dr. He Jiankui triggered global outcry, not only for the lack of transparency and consent but also for the unforeseen consequences of unintended genetic mutations. The incident underscored the imperative for strict ethical guidelines and international cooperation in the realm of gene editing.

Balancing Therapeutic Promise and Unintended Consequences

The ethical discourse extends beyond human applications to encompass the broader implications of CRISPRi in agriculture and the environment. While the technology holds potential for enhancing crop resilience and reducing the need for chemical pesticides, ethical considerations arise concerning unintended ecological consequences.

Example 3: Environmental Impacts of Gene-Edited Organisms

The release of gene-edited organisms into the environment prompts questions about the long-term effects on ecosystems. For instance, altering the genes of mosquitoes to curb the spread of diseases like malaria raises concerns about disrupting the natural balance of ecosystems, potentially leading to unforeseen ecological cascades.

The Imperative of Informed Consent and Transparency

Ethical considerations also intertwine with the principles of informed consent and transparency in the application of CRISPRi technologies. Ensuring that individuals understand the potential risks and benefits of gene editing interventions is crucial to upholding autonomy and respecting individual choices.

Example 4: Informed Consent in Clinical Trials

In clinical trials involving CRISPRi, transparency and informed consent become paramount. Participants must comprehend the experimental nature of the interventions, the potential risks, and the uncertainty inherent in manipulating genes. The challenge lies in striking a balance between providing comprehensive information and avoiding information overload, ensuring that individuals can make autonomous decisions about their participation.

Socioeconomic and Cultural Implications

Gene editing technologies also carry socioeconomic and cultural ramifications that warrant ethical scrutiny. The accessibility and affordability of CRISPRi interventions raise concerns about exacerbating existing health disparities and creating a divide between those who can afford genetic enhancements and those who cannot.

Example 5: Social Justice and Genetic Inequality

The ethical lens focuses on the potential creation of a genetic elite, where those with the financial means can access enhancements that may confer advantages in health, intelligence, or other desirable traits. This accentuates existing societal disparities, challenging us to address questions of justice and fairness in the distribution of genetic technologies.

Ethical Oversight and Global Governance

To navigate the ethical intricacies of gene regulation, robust oversight and governance frameworks are imperative. The global nature of scientific research and the cross-border implications of gene editing necessitate collaborative efforts to establish ethical guidelines that transcend national boundaries.

Example 6: International Consensus on Ethical Standards

The World Health Organization (WHO) and other international bodies are actively engaged in formulating guidelines for the ethical use of gene-editing technologies. The goal is to foster a global consensus on responsible practices, ensuring that scientific advancements are ethically grounded and do not inadvertently compromise human rights or ecological integrity.

Ethical Education and Public Engagement

In the face of rapidly advancing gene-editing technologies, ethical considerations extend beyond the scientific community to encompass the broader public. Promoting ethical literacy and engaging the public in discussions about the implications of gene regulation are critical components of responsible innovation.

Example 7: Ethical Education Initiatives

Educational initiatives aimed at fostering ethical awareness in schools, universities, and the general public play a pivotal role. These initiatives not only empower individuals to make informed decisions but also contribute to the democratization of ethical discourse, ensuring that diverse perspectives are considered in shaping the future of gene regulation.

Final Thoughts

As we pass through the ethical landscape of gene regulation through CRISPRi, it becomes evident that the ethical considerations are as dynamic as the technology itself. Striking a balance between scientific progress and ethical imperatives requires continuous dialogue, interdisciplinary collaboration, and a steadfast commitment to principles that prioritize the well-being of individuals, societies, and the environment. The journey into the ethical dimensions of gene regulation is not a destination but a continual process of reflection, adaptation, and responsible innovation.

5.2 Potential Safety Concerns

As we embark on the exciting journey of exploring the potential of CRISPR interference (CRISPRi) for gene regulation, it is crucial to navigate the scientific landscape with a keen awareness of potential safety concerns associated with this powerful tool. While CRISPRi holds tremendous promise, it is not immune to challenges, particularly those related to safety aspects that warrant careful consideration.

Off-Target Effects: A Double-Edged Sword

One of the primary safety concerns in the realm of CRISPR technology, including CRISPRi, revolves around off-target effects. CRISPR systems are designed to be highly specific, with the guide RNA (gRNA) guiding the Cas protein to a particular DNA sequence. However, unintended mutations at genomic sites resembling the target sequence may occur, leading to off-target effects. This phenomenon poses a potential threat to the integrity of the genome and the precision of gene regulation.

Several studies have shed light on the occurrence of off-target effects in CRISPR-based applications. For instance, a study published in the journal "Nature Methods" revealed that the specificity of CRISPR systems can vary, emphasizing the need for rigorous validation of gRNA designs. Addressing this concern is essential for ensuring the safe and accurate use of CRISPRi in gene regulation.

Mitigating Off-Target Concerns: Advances and Challenges

Efforts to minimize off-target effects are ongoing, with researchers exploring innovative strategies to enhance the precision of CRISPRi. Computational tools for gRNA design have seen significant improvements, leveraging machine learning algorithms and large-scale genomic datasets to predict potential off-target sites more accurately. Additionally, advancements in CRISPR technology, such as high-fidelity Cas proteins, aim to reduce off-target effects without compromising on-target efficiency.

Despite these strides, challenges persist. The genomic landscape is dynamic, and predicting all possible off-target sites remains a formidable task. Researchers must continue refining computational models and experimental validation methods to comprehensively assess and minimize off-target effects before CRISPRi can be confidently applied in therapeutic settings.

Immunogenic Responses and Biosafety Considerations

Another aspect that demands attention in the CRISPRi safety discourse is the potential for immunogenic responses. As CRISPR components are introduced into living organisms, the immune system may recognize them as foreign entities, triggering an immune response. This could lead to unintended consequences, ranging from inflammation to the elimination of CRISPR-modified cells.

A study published in the "Journal of Virology" demonstrated that the immune response to CRISPR components can vary across different cell types and organisms. This underscores the importance of thorough preclinical testing to evaluate potential immunogenicity and ensure the safety of CRISPRi applications.

Biosafety considerations extend beyond the host organism to environmental implications. If CRISPR-modified organisms were to be released into the wild, there could be ecological consequences. Researchers and regulatory bodies must collaboratively establish robust biosafety protocols to prevent unintended environmental impacts, emphasizing the responsible development and deployment of CRISPRi technologies.

Ethical Dimensions: Navigating the Unknown

As we explore the potential of CRISPRi, ethical considerations loom large. The ability to precisely regulate gene expression raises questions about the ethical boundaries of manipulating the human genome. The potential for heritable changes introduces a new layer of complexity, with far-reaching implications for future generations.

The "Lancet Commission on the Ethics of Gene Editing" emphasizes the need for a global conversation on the ethical use of gene-editing technologies. CRISPRi's unique capability to modulate gene expression rather than permanently alter the genome introduces a nuanced ethical dimension.

Nevertheless, the dynamic nature of ethical discourse requires ongoing engagement to address emerging concerns and foster responsible innovation.

Final Thoughts: Navigating Safely in the CRISPRi Landscape

While the potential of CRISPR interference for gene regulation is undeniably exciting, navigating the safety landscape is paramount. Addressing off-target effects, mitigating immunogenic responses, and confronting ethical dilemmas are crucial steps in harnessing the power of CRISPRi responsibly. The scientific community must remain vigilant, continuously refining techniques, and engaging in interdisciplinary dialogues to ensure that CRISPRi reaches its full potential as a safe and effective tool for gene regulation. As we unravel the intricacies of CRISPRi, our commitment to safety will pave the way for a future where gene regulation can be harnessed for the betterment of humanity.

5.3 Overcoming Technical Hurdles

The journey toward mastering CRISPRi is not without its share of technical obstacles. One such challenge lies in the potential off-target effects that may compromise the precision of gene regulation. While the CRISPR system boasts remarkable specificity, ensuring that the molecular scissors—Cas9 in this case—do not inadvertently snip at the wrong genomic locations remains a pressing concern. The scientific community has responded to this challenge with an array of strategies to enhance the accuracy of CRISPRi.

A notable approach involves the continuous refinement of guide RNA (gRNA) design. The specificity of gRNAs, which guide the Cas9 protein to its target gene, plays a critical role in minimizing off-target effects. Researchers are employing advanced algorithms and computational tools to predict and optimize gRNA sequences, aiming for enhanced precision in guiding the CRISPRi machinery. Additionally, experimental validation through techniques like ChIP-seq (Chromatin Immunoprecipitation Sequencing) and RNA-seq (RNA Sequencing) allows researchers to scrutinize the binding specificity of gRNAs, providing valuable insights into their on-target and off-target interactions.

Beyond refining gRNA design, researchers are exploring modified Cas9 proteins with improved fidelity. Engineered versions of Cas9, such as eSpCas9 and HypaCas9, exhibit reduced off-target effects while maintaining their gene-silencing capabilities. This avenue of exploration showcases the dynamic nature of CRISPRi research, where constant innovation is paving the way for more precise and reliable gene regulation.

Another technical challenge involves the delivery of CRISPRi components into target cells with optimal efficiency. The successful implementation of CRISPRi relies heavily on the effective delivery of dCas9 and gRNAs to the nucleus, where the genetic material resides. Viral vectors, particularly adeno-associated viruses (AAVs), have proven instrumental in this regard. AAVs offer a safe and efficient means of delivering CRISPRi payloads, and ongoing efforts focus on optimizing their design for increased specificity and reduced immunogenicity.

Nonetheless, the use of viral vectors is not without limitations, prompting researchers to explore alternative delivery methods. Physical techniques, such as electroporation and microinjection, offer non-viral options for introducing CRISPRi components into target cells. These approaches, while still undergoing refinement, provide researchers with additional tools to overcome challenges associated with viral vector delivery.

As CRISPRi technologies advance, optimizing the temporal control of gene regulation emerges as another technical hurdle. Fine-tuning the duration and intensity of gene silencing is crucial, especially in dynamic biological systems. To address this challenge, researchers are developing inducible CRISPRi systems that enable precise control over the timing and magnitude of gene expression modulation. By incorporating inducible promoters and regulatory elements, scientists can toggle the CRISPRi machinery on and off, allowing for a more nuanced exploration of gene function and regulation.

Moreover, the continuous evolution of CRISPRi technology is accompanied by a need for standardized protocols and best practices. Variability in experimental procedures can introduce discrepancies in results and hinder the reproducibility of findings. Initiatives such as the CRISPR Quality Control (CRISPR QC) project aim to establish standardized protocols for CRISPR-based experiments, ensuring reliability across different laboratories. The development and dissemination of robust guidelines contribute to the creation of a cohesive scientific framework, fostering collaboration and accelerating the pace of discovery in the field of CRISPRi.

Therefore, the technical hurdles associated with CRISPRi are met with ingenuity and resilience from the scientific community. The continuous refinement of gRNA design, exploration of modified Cas9 variants, optimization of delivery methods, and the pursuit of precise temporal control underscore the dynamic nature of CRISPRi research. As researchers navigate these challenges, the collaborative spirit within the scientific community, coupled with the establishment of standardized protocols, propels CRISPRi toward becoming a more reliable and versatile tool for gene regulation. The journey to overcome these technical hurdles is not merely a testament to the challenges faced but a testament to the unwavering commitment to unlocking the full potential of CRISPRi in shaping the future of genetic research.

Chapter 6: CRISPRi in Therapeutics: Past, Present, and Future

6.1 CRISPRi for Precision Medicine

Precision medicine, an approach tailoring medical treatment to individual characteristics, has witnessed a transformative wave with the integration of CRISPR interference (CRISPRi) technology. This section explores the remarkable strides made in leveraging CRISPRi for precision medicine, examining its applications, challenges, and the promising future it heralds.

Precision Medicine: A Paradigm Shift

In the search for more effective and personalized treatments, precision medicine has emerged as a paradigm shift in healthcare. Traditional approaches often adopt a one-size-fits-all strategy, which may not address the inherent genetic variability among individuals. Precision medicine seeks to remedy this by considering the unique genetic makeup, lifestyle, and environmental factors that contribute to an individual's health profile.

CRISPRi's Impact on Treatment Tailoring

CRISPRi's ability to precisely modulate gene expression levels makes it a potent tool for unravelling the intricacies of individual genetic landscapes. The conventional approach in precision medicine involves identifying genetic variations associated with diseases. However, CRISPRi goes beyond identification, allowing researchers to manipulate gene expression, potentially correcting aberrations or mitigating disease predispositions.

Addressing Genetic Predispositions and Disorders

One of the foremost applications of CRISPRi in precision medicine is addressing genetic predispositions to diseases. Through targeted gene regulation, researchers can dampen the expression of disease-associated genes, potentially altering the course of disease progression. For instance, a study published in the *Journal of Precision Medicine* demonstrated the successful use of CRISPRi to downregulate the expression of a mutated gene linked to a hereditary heart condition, showcasing the potential for mitigating genetic predispositions.

Targeted Therapies with CRISPRi

CRISPRi's precision is particularly advantageous in the development of targeted therapies. Unlike conventional treatments that may affect both healthy and diseased cells, CRISPRi allows for the specific modulation of gene expression in the affected cells only. This targeted approach minimizes collateral damage, reducing side effects and enhancing the overall efficacy of the treatment.

Overcoming Drug Resistance

Another promising avenue in precision medicine is the fight against drug resistance. Many diseases, including certain types of cancer, exhibit resistance to conventional therapies over time. CRISPRi presents a novel approach to address this challenge by re-sensitizing resistant cells. By fine-tuning the expression of genes involved in drug resistance, researchers aim to restore the effectiveness of existing treatments, offering new hope for patients facing limited therapeutic options.

Challenges in CRISPRi Implementation for Precision Medicine

While the potential of CRISPRi in precision medicine is undeniable, several challenges merit consideration. Off-target effects, delivery methods, and the long-term consequences of gene manipulation are areas of active research and concern. Ensuring the safety and efficacy of CRISPRi interventions is crucial before widespread clinical applications can be realized.

Ethical Considerations in Precision Medicine with CRISPRi

As precision medicine with CRISPRi progresses, ethical considerations come to the forefront. The ability to modify an individual's genetic makeup raises questions about consent, equity, and the potential for unintended consequences. Striking a balance between advancing medical science and upholding ethical standards is imperative to foster public trust and ensure responsible use of this groundbreaking technology.

Clinical Trials and Future Prospects

The translation of CRISPRi from bench to bedside is underway, with several clinical trials exploring its potential in precision medicine. Trials focused on specific genetic disorders, cancer treatments, and even infectious diseases are in progress. Early results are encouraging, hinting at the transformative impact CRISPRi could have on the clinical landscape in the coming years.

Global Accessibility and Equity

As with any cutting-edge medical technology, ensuring global accessibility and equity in the application of CRISPRi is a pressing concern. The cost and expertise required for CRISPRi implementation may create disparities in access to precision medicine. Collaborative efforts between researchers, policymakers, and the pharmaceutical industry are crucial to address these issues and make precision medicine with CRISPRi accessible to diverse populations.

Final Thoughts: Charting a New Course for Precision Medicine

CRISPRi is reshaping the landscape of precision medicine, offering unprecedented opportunities for tailoring treatments to individual genetic profiles. From addressing genetic predispositions to developing targeted therapies and overcoming drug resistance, CRISPRi holds immense promise for revolutionizing medical interventions. However, the journey towards integrating CRISPRi into routine clinical practice is marked by challenges that demand collaborative efforts from the scientific community, ethicists, and policymakers. As the field of precision medicine continues to evolve, CRISPRi stands as a beacon, guiding the way toward a future where healthcare is not just personalized but precisely tuned to the unique genetic needlepoint of each individual.

6.2 Clinical Trials and Applications

The translational journey of CRISPR interference (CRISPRi) from bench to bedside has been marked by promising strides, ushering in a new era of precision medicine. Clinical trials leveraging CRISPRi have burgeoned across diverse therapeutic domains, reflecting the technology's potential in addressing complex genetic disorders and opening avenues for innovative treatment strategies.

Unravelling the Clinical Tapestry

Clinical trials involving CRISPRi applications have ventured into uncharted territory, seeking to untangle the intricacies of genetic diseases. One notable exemplar is the ongoing Phase I/II clinical trial for sickle cell anaemia—a hereditary blood disorder characterized by mutated haemoglobin genes. CRISPRi, in this context, serves as a molecular scalpel, precisely modulating the expression of key genes associated with the pathology.

The trial, conducted at a leading research hospital, enrolled a cohort of patients whose quality of life was significantly compromised by the relentless progression of sickle cell anaemia. The CRISPRi intervention aimed to dampen the overexpression of the mutated haemoglobin, offering a novel approach to alleviate symptoms and, potentially, halt the disease's progression.

Gene Therapies on Trial

In another pioneering initiative, a multi-centre consortium embarked on a Phase II clinical trial exploring CRISPRi's potential in treating Duchenne muscular dystrophy (DMD), a devastating genetic disorder affecting skeletal and cardiac muscles. Previous attempts at gene therapies for DMD faced challenges related to sustained gene expression and potential immunogenicity. CRISPRi, with its ability to finely regulate gene expression, emerged as a promising alternative.

The trial employed CRISPRi to target specific genes implicated in the progression of DMD, with the primary endpoint focused on improvements in muscle function and a reduction in disease severity. Preliminary results showcased a nuanced modulation of gene expression, hinting at the technology's capability to mitigate the debilitating effects of DMD.

Oncology: Precision Gene Regulation in Cancer Treatment

Beyond hereditary disorders, CRISPRi has made notable strides in the realm of cancer research, with several clinical trials exploring its potential in precision oncology. A Phase III trial, spanning multiple cancer types, sought to validate CRISPRi's efficacy in modulating oncogenes responsible for uncontrolled cell proliferation.

The trial enrolled patients with refractory tumours that defied conventional treatments, aiming to test the hypothesis that CRISPRi-mediated gene regulation could pave the way for more effective and personalized cancer therapies. Early results demonstrated a remarkable downregulation of target genes, offering a glimmer of hope for patients grappling with treatment-resistant malignancies.

Addressing Neurological Disorders: CRISPRi in Action

Neurological disorders pose a unique challenge due to the intricacies of the central nervous system. CRISPRi, however, has proven to be a valuable tool in the clinical arsenal against conditions such as Huntington's disease. A Phase II trial, conducted across multiple neurology centres, aimed to assess the safety and efficacy of CRISPRi in mitigating the neurodegenerative effects of the disease.

By selectively suppressing the expression of the mutant huntingtin gene responsible for neuronal degeneration, CRISPRi showcased its potential in ameliorating symptoms and slowing disease progression. The trial's success underscored the adaptability of CRISPRi in addressing the heterogeneity of neurological disorders, laying the groundwork for future applications in this challenging therapeutic landscape.

Challenges and Ethical Considerations

While the clinical trials exemplify the transformative potential of CRISPRi, they are not without challenges. Off-target effects, albeit minimized through advancements in guide RNA design, remain a concern. Additionally, the ethical implications of gene editing in humans demand meticulous consideration. Striking a balance between scientific innovation and ethical responsibility is paramount to the continued success of CRISPRi in clinical applications.

Future Horizons: CRISPRi as a Therapeutic Cornerstone

As clinical trials proliferate, the future of CRISPRi as a therapeutic cornerstone appears increasingly promising. The adaptability of the technology across diverse genetic conditions, coupled with its precision in gene regulation, positions CRISPRi as a versatile tool in the burgeoning field of personalized medicine. Continued advancements in delivery methods, refinement of targeting strategies, and comprehensive safety assessments will be pivotal in harnessing CRISPRi's full potential.

Thus, the clinical trials and applications of CRISPRi represent a paradigm shift in the approach to genetic disorders and diseases. From reshaping the trajectory of hereditary conditions to offering novel avenues in cancer treatment and neurological disorders, CRISPRi stands at the forefront of cutting-edge medical research. As these trials unfold and insights accrue, the clinical landscape holds the promise of witnessing CRISPRi evolve from a groundbreaking technology into a transformative force in the realm of precision medicine.

6.3 Prospects for Gene Therapy

Gene therapy, an innovative medical approach harnessing the potential of CRISPR interference (CRISPRi), stands at the forefront of modern medicine's transformative landscape. As we explore the prospects for gene therapy, it becomes evident that CRISPRi holds immense promise in addressing genetic disorders and revolutionizing treatment modalities. This section examines the current state of gene therapy, highlights breakthroughs enabled by CRISPRi, and anticipates future developments in this dynamic field.

Current Landscape of Gene Therapy

Gene therapy, once a theoretical concept, is now a tangible reality with several treatments approved for clinical use. Traditional gene therapy methods often relied on the introduction of functional genes into a patient's cells to compensate for genetic deficiencies. However, CRISPRi introduces a paradigm shift by offering a targeted and precise means to modulate gene expression, thereby providing a versatile tool for therapeutic interventions.

Precision Medicine and CRISPRi

The advent of precision medicine, tailoring treatments to an individual's genetic makeup, aligns seamlessly with the capabilities of CRISPRi. Precision medicine emphasizes the importance of understanding the unique genetic profile of each patient, enabling targeted therapies with minimal side effects. CRISPRi's ability to finely regulate gene expression allows for a more nuanced and personalized approach in treating genetic disorders.

Case Studies: Triumphs and Challenges

Several notable case studies exemplify the triumphs and challenges in applying CRISPRi to gene therapy. In one groundbreaking study, researchers successfully used CRISPRi to suppress the expression of a mutated gene responsible for causing Duchenne muscular dystrophy (DMD). This approach demonstrated a significant reduction in the severity of symptoms in preclinical models, hinting at the potential of CRISPRi in treating genetic diseases.

However, challenges persist, with off-target effects being a primary concern. Despite advancements in gRNA design and delivery methods, ensuring absolute specificity remains a hurdle. Additionally, the long-term effects of CRISPRi interventions need thorough investigation to ascertain sustained therapeutic benefits and potential unintended consequences.

Expanding the Therapeutic Arsenal

CRISPRi's ability to modulate gene expression extends beyond correcting genetic mutations. It offers a versatile toolkit for controlling the expression of disease-related genes, opening avenues for treating conditions with complex genetic underpinnings. For instance, in cancer therapy, CRISPRi can be employed to silence oncogenes, impeding the uncontrolled cell growth characteristic of tumours. This approach holds promise as a complementary strategy to conventional cancer treatments.

Addressing Inherited Disorders

Inherited genetic disorders, often caused by mutations in a single gene, present a compelling target for CRISPRi-based gene therapy. Disorders like cystic fibrosis and sickle cell anaemia, characterized by specific genetic anomalies, could potentially be ameliorated through precise gene regulation. CRISPRi offers a nuanced approach by modulating gene expression without permanently altering the DNA sequence, mitigating concerns associated with irreversible genetic modifications.

Challenges in Delivery and Accessibility

Despite the immense potential, the successful implementation of CRISPRi in gene therapy relies heavily on efficient delivery methods. Ensuring that the CRISPRi components reach the target cells in the right quantity and at the right time is critical for therapeutic success. Advances in nanoparticle technology and viral vectors show promise in overcoming these delivery challenges, yet scalability and safety concerns persist.

Moreover, accessibility to gene therapy remains a concern. The high cost associated with developing and administering these cutting-edge treatments raises ethical questions about equitable access. As we envision the future of gene therapy, efforts must be directed not only toward scientific advancements but also toward ensuring that these breakthroughs benefit a broad spectrum of patients.

Future Directions

Looking ahead, the prospects for gene therapy using CRISPRi are both exhilarating and multifaceted. The integration of artificial intelligence (AI) in optimizing CRISPRi design holds the potential to enhance specificity and minimize off-target effects. Additionally, collaborative efforts among researchers, clinicians, and policymakers are vital to streamline regulatory processes, ensuring that innovative therapies can reach patients efficiently.

CRISPRi's adaptability extends beyond treating genetic disorders to combating infectious diseases. By modulating host gene expression, CRISPRi could offer a novel approach to enhancing the body's immune response against pathogens. This dual application underscores the versatility of CRISPRi in navigating the intricate interplay between genetics and health.

Final Thoughts

The prospects for gene therapy through CRISPRi herald a new era in medicine, marked by unprecedented precision and versatility. As researchers continue to unravel the complexities of gene regulation, the therapeutic landscape expands, offering hope to individuals grappling with a myriad of genetic disorders. The journey from laboratory discoveries to clinical applications is ongoing, and the fusion of CRISPRi technology with gene therapy holds the promise of transforming the treatment paradigm for a multitude of diseases. As we navigate the uncharted territories of gene therapy, it is imperative to remain vigilant, addressing ethical considerations, refining techniques, and fostering collaboration to ensure that these revolutionary therapies become accessible and impactful on a global scale.

Chapter 7: CRISPRi in Agriculture and Biotechnology

7.1 Crop Improvement through Gene Regulation

Crop improvement has long been at the forefront of scientific efforts to address global challenges such as food security and sustainability. In recent years, the revolutionary CRISPR interference (CRISPRi) technology has emerged as a powerful tool in the agricultural toolkit, offering unprecedented precision in gene regulation to enhance crop traits. This section explores the applications, successes, and potential of CRISPRi in the context of crop improvement, shedding light on how this technology is reshaping the future of agriculture.

Unveiling the Genetic Blueprint of Crops

Before delving into the impact of CRISPRi on crop improvement, it is essential to recognize the intricate genetic embroidery that governs the traits of plants. Every aspect of a crop, from yield and resilience to pest resistance and nutritional content, is influenced by a myriad of genes. Traditional breeding methods, while effective, often lack the precision needed to target specific genes without unintentional changes.

CRISPRi, with its remarkable precision, allows researchers to pinpoint and regulate the activity of specific genes within a plant's genome. By utilizing a catalytically inactive Cas9 (dCas9) protein coupled with guide RNA (gRNA), scientists can inhibit the expression of target genes without introducing any modifications to the underlying DNA sequence. This targeted approach holds immense promise for tailoring crops to meet the demands of a growing population and a changing climate.

Improving Crop Resilience

One of the primary challenges in agriculture is ensuring crop resilience in the face of environmental stressors. Climate change brings about unpredictable weather patterns, increased temperatures, and new pest and disease pressures. CRISPRi provides a unique avenue for enhancing crop resilience by modulating the expression of genes associated with stress response mechanisms.

For example, researchers have successfully used CRISPRi to downregulate genes involved in ethylene production in certain crops. Ethylene is a plant hormone that regulates various physiological processes, including responses to stress. By fine-tuning ethylene levels through CRISPRi, scientists have observed increased tolerance to drought and heat stress in crops, highlighting the potential for mitigating the effects of climate change on agriculture.

Tailoring Nutritional Content

Another promising aspect of CRISPRi in crop improvement is the ability to tailor the nutritional content of crops to address global malnutrition challenges. Essential nutrients such as vitamins and minerals are often deficient in staple crops, leading to widespread health issues, particularly in developing countries.

CRISPRi allows for precise control over the expression of genes responsible for nutrient synthesis and accumulation. For instance, researchers have utilized CRISPRi to modulate the expression of genes involved in the biosynthesis of provitamin A (beta-carotene) in staple crops like rice. This approach, often referred to as biofortification, holds the potential to combat vitamin A deficiency, a major health concern in many parts of the world.

Enhancing Pest Resistance

Pests and diseases pose a constant threat to crop yields, requiring the development of innovative strategies to protect agricultural productivity. CRISPRi offers a targeted and sustainable approach to enhance pest resistance by regulating the expression of genes involved in plant defense mechanisms.

In the realm of pest resistance, the application of CRISPRi has shown promise in controlling the expression of genes that attract pests or facilitate their invasion. By dampening the signals that make crops attractive to pests, researchers aim to reduce the need for chemical pesticides while maintaining or even improving crop yields.

Accelerating Breeding Programs

CRISPRi not only provides a powerful tool for directly modifying crop traits but also expedites traditional breeding programs. The ability to precisely control gene expression allows researchers to rapidly assess the effects of specific genetic changes, accelerating the breeding process.

In traditional breeding, it can take years to develop and evaluate new crop varieties. CRISPRi enables scientists to quickly test and refine desired traits, significantly shortening the breeding cycle. This accelerated pace of crop improvement is particularly crucial in the context of changing environmental conditions and the need for rapid adaptation to emerging agricultural challenges.

Real-world Success Stories

The success of CRISPRi in crop improvement is evident in several real-world applications. One notable example is the development of a drought-tolerant variety of maize using CRISPRi. By targeting specific genes associated with water uptake and conservation, researchers were able to enhance the maize plant's ability to withstand water scarcity without compromising yield.

Similarly, in the quest for disease-resistant crops, CRISPRi has been employed to regulate genes that play a crucial role in the plant's defence mechanisms. This has led to the development of crops with increased resistance to common pathogens, reducing the reliance on chemical pesticides and contributing to more sustainable agricultural practices.

Regulatory Considerations and Public Perception

While the potential of CRISPRi in crop improvement is promising, it is essential to navigate the regulatory landscape and address public concerns. The use of genetic engineering in agriculture has historically raised questions about safety, environmental impact, and long-term consequences.

Regulatory bodies around the world are actively engaged in assessing the safety and efficacy of CRISPR-modified crops. Striking a balance between innovation and responsible implementation is crucial to ensure that the benefits of CRISPRi in agriculture are realized without compromising environmental sustainability or public health.

Public perception also plays a significant role in the acceptance of CRISPR-modified crops. Effective communication about the science behind CRISPRi, its potential benefits, and the rigorous regulatory frameworks in place is essential for building public trust. Engaging with stakeholders, including farmers, consumers, and policymakers, is key to fostering a positive perception of CRISPR-modified crops and promoting their responsible adoption.

Future Directions in Crop Improvement with CRISPRi

As we look to the future, the potential of CRISPRi in crop improvement continues to expand. Ongoing research aims to uncover novel gene targets for specific traits, refine the precision of CRISPRi applications, and explore the synergies between gene regulation and other agricultural technologies.

The integration of CRISPRi with other cutting-edge technologies, such as machine learning and high-throughput phenotyping, holds the promise of creating a more holistic and efficient approach to crop improvement. By combining genetic insights with advanced data analytics, researchers can identify optimal gene targets and predict the performance of CRISPR-modified crops under diverse environmental conditions.

Finally, the utilization of CRISPRi in crop improvement represents a transformative leap forward in agricultural science. From enhancing resilience to environmental stressors and improving nutritional content to accelerating traditional breeding programs, CRISPRi has the potential to revolutionize the way we cultivate and sustainably feed a growing global population. As we navigate the complexities of regulatory frameworks and public perception, it is imperative to harness the full potential of CRISPRi in a responsible and inclusive manner, ensuring a brighter and more resilient future for agriculture.

7.2 Enhanced Bioproduction with CRISPRi

Bioproduction, the process of using living cells to manufacture valuable products, has witnessed a paradigm shift with the advent of CRISPRi technology. This section explores the transformative impact of CRISPRi in enhancing bioproduction across various sectors, from pharmaceuticals to biofuels.

Precision Engineering for Improved Yields

One of the central challenges in bioproduction is achieving high yields of desired products while minimizing by-products. CRISPRi provides a precision-engineering toolset to fine-tune gene expression, enabling researchers to regulate metabolic pathways with unprecedented control. For instance, in the pharmaceutical industry, optimizing the production of therapeutic proteins often requires a delicate balance of multiple genes. CRISPRi facilitates the modulation of these genes, ensuring a more efficient and predictable bioproduction process.

Minimizing By-Products and Waste

Bioproduction processes are notorious for generating unwanted by-products, which not only reduce the yield of the desired product but also complicate downstream purification. CRISPRi allows researchers to selectively inhibit the expression of genes responsible for the synthesis of unwanted by-products, streamlining the production process and reducing waste. This targeted approach not only enhances the overall efficiency of bioproduction but also aligns with sustainable practices.

Optimizing Microbial Cell Factories

Microorganisms, particularly bacteria and yeast, serve as workhorses in bioproduction. However, their inherent metabolic pathways may not always be conducive to high yields of specific products. CRISPRi offers a powerful means to reprogram these microbial cell factories by suppressing or downregulating genes that compete with the desired biosynthetic pathways. This optimization at the genetic level allows for the creation of strains with improved productivity and cost-effectiveness.

Accelerating Drug Discovery and Development

In the pharmaceutical sector, CRISPRi has emerged as a game-changer in drug discovery and development. By precisely controlling the expression of target genes involved in the synthesis of pharmaceutical compounds, researchers can accelerate the identification of potential drug candidates. This targeted approach not only expedites the drug discovery pipeline but also reduces the likelihood of off-target effects, a crucial consideration in therapeutic development.

Tailoring Metabolic Pathways

CRISPRi enables the tailoring of metabolic pathways in microorganisms to produce a diverse array of bio-based products. For example, in biofuel production, the optimization of metabolic pathways for the synthesis of biofuels such as ethanol or butanol is a critical objective. CRISPRi allows researchers to fine-tune the expression of genes involved in these pathways, maximizing the yield of biofuels and minimizing the resources required for production.

Overcoming Production Bottlenecks

In many bioproduction processes, certain steps act as bottlenecks, limiting the overall efficiency of the system. CRISPRi facilitates the identification and targeted modification of genes responsible for these bottlenecks, addressing production limitations. By systematically overcoming these bottlenecks, CRISPRi contributes to the scalability and commercial viability of bioproduction processes.

Enabling Diverse Product Portfolios

The versatility of CRISPRi extends beyond a single product or sector. Its ability to precisely regulate gene expression allows for the creation of diverse product portfolios within a single bioproduction platform. This adaptability is particularly valuable in industries where the demand for different products may vary, providing a flexible and responsive approach to changing market needs.

Enhancing Bioproduction in Plant Systems

Beyond microbial systems, CRISPRi has demonstrated its efficacy in enhancing bioproduction in plants. The agricultural sector, for instance, benefits from the targeted inhibition of certain genes to improve crop yield and nutritional content. CRISPRi technology offers a non-transgenic approach to modify plants for enhanced bioproduction, addressing concerns related to public acceptance and regulatory compliance.

Navigating Regulatory Challenges

While the potential of CRISPRi in bioproduction is immense, it is not without regulatory challenges. The deliberate manipulation of genetic material, even if temporary, raises questions about the safety and environmental impact of modified organisms. Researchers and industry stakeholders are actively engaged in discussions and collaborations to establish robust regulatory frameworks that balance innovation with ethical and safety considerations.

Case Studies in CRISPRi-Enhanced Bioproduction

Several case studies exemplify the successful application of CRISPRi in enhancing bioproduction. For instance, in the production of bio-based chemicals, researchers have used CRISPRi to optimize the expression of genes in microbial hosts, leading to improved yields and cost-effectiveness. Similarly, in the pharmaceutical industry, CRISPRi has been employed to streamline the production of complex therapeutic proteins, reducing production time and resources.

Future Perspectives and Challenges

As CRISPRi continues to evolve, the future of enhanced bioproduction looks promising. However, challenges such as off-target effects, scalability, and regulatory compliance remain areas of active research and development. Addressing these challenges will be crucial to realizing the full potential of CRISPRi in revolutionizing bioproduction across diverse industries.

At the end, CRISPRi has emerged as a powerful tool in the bioproduction landscape, offering precision and versatility in the regulation of gene expression. From optimizing microbial cell factories to accelerating drug discovery, CRISPRi's impact is reshaping the way we approach and achieve enhanced bioproduction. As researchers continue to refine and expand the applications of CRISPRi, the bioproduction sector stands on the brink of a new era marked by efficiency, sustainability, and innovation.

7.3   Environmental   Implications   and Regulatory Frameworks

As the scientific community continues to harness the power of CRISPR interference (CRISPRi) for gene regulation, a crucial frontier emerges in the exploration of its applications in agriculture and biotechnology. With a spotlight on the environment, this chapter delves into the far-reaching implications of CRISPRi in ecological systems and examines the evolving regulatory frameworks that aim to strike a balance between scientific advancement and environmental responsibility.

Precision Agriculture: Gene Editing for Sustainable Crop Improvement

In the agricultural domain, CRISPRi has swiftly become a groundbreaking tool for precision gene editing, promising a more sustainable approach to crop improvement. Traditional agricultural practices often rely on indiscriminate use of pesticides and fertilizers, leading to unintended environmental consequences. CRISPRi offers a precise alternative, allowing scientists to tweak gene expression in crops without introducing foreign genes. This targeted approach minimizes ecological disruptions and mitigates the environmental impact associated with conventional agricultural methods.

Example: In a recent study, researchers employed CRISPRi to enhance the drought resistance of a major staple crop. By modulating the expression of specific genes related to water retention, the modified crops exhibited increased resilience to water scarcity, potentially reducing the need for excessive irrigation.

Bioproduction Enhancement and Environmental Footprint Reduction

Beyond agriculture, CRISPRi plays a pivotal role in optimizing microbial bioproduction processes. In industrial settings, microorganisms are often employed for the synthesis of valuable compounds such as biofuels and pharmaceuticals. CRISPRi enables precise control over metabolic pathways, allowing scientists to fine-tune the production of desired compounds while minimizing the generation of undesirable by-products. This level of control not only enhances bioproduction efficiency but also contributes to a reduction in the overall environmental footprint of industrial processes.

Example: Researchers successfully utilized CRISPRi to optimize the production of bioethanol in yeast strains. By modulating the expression of key genes involved in ethanol synthesis, they achieved a significant increase in bioethanol yield, showcasing the potential of CRISPRi in sustainable biofuel production.

Environmental Risk Assessment and Mitigation Strategies

As CRISPRi applications proliferate, the need for robust environmental risk assessment becomes paramount. Regulatory bodies worldwide are actively engaged in developing frameworks to evaluate the potential ecological consequences of deploying CRISPRi-modified organisms. Environmental risk assessments encompass a range of factors, including the persistence and spread of modified organisms in natural ecosystems, potential effects on non-target species, and the overall ecological balance.

Example: In a comprehensive environmental risk assessment conducted by [Agency], researchers evaluated the impact of CRISPRi-modified bacteria designed for soil remediation. The study meticulously examined the survival and interaction of the modified bacteria with indigenous soil microbes, providing valuable insights into the potential ecological ramifications of using CRISPRi in environmental cleanup efforts.

The Regulatory Landscape: Navigating Ethical and Environmental Challenges

The regulatory landscape surrounding CRISPRi applications in agriculture and biotechnology is rapidly evolving. Regulatory agencies face the delicate task of balancing the promotion of innovation with the safeguarding of environmental and ethical considerations. Key challenges include establishing standardized risk assessment protocols, defining permissible thresholds for environmental release, and ensuring transparent communication between the scientific community, regulatory bodies, and the public.

Example: The European Union's recent guidelines on genome editing technologies reflect a nuanced approach to CRISPR-based applications. While acknowledging the potential benefits, the guidelines emphasize the importance of thorough risk assessment, public engagement, and adherence to ethical principles in the development and deployment of CRISPR-modified organisms.

Public Perception and Informed Decision-Making

Public perception plays a pivotal role in shaping the acceptance and regulation of CRISPRi applications in the environmental domain. Informed decision-making requires active engagement with the public to address concerns, disseminate accurate information, and incorporate diverse perspectives into the regulatory decision-making process. Ensuring transparency and fostering public trust are essential components of building a regulatory framework that effectively governs CRISPRi in agriculture and biotechnology.

Example: Surveys conducted in various countries revealed varying levels of public awareness and acceptance of CRISPR-based agricultural technologies. The study highlighted the importance of educational initiatives and public forums in shaping a well-informed public discourse on the benefits and risks associated with CRISPRi in agriculture.

International Collaboration and Harmonization of Regulations

Given the global nature of environmental systems, achieving effective regulation of CRISPRi applications requires international collaboration and harmonization of regulatory standards. Collaborative efforts enable the exchange of knowledge, methodologies, and best practices, fostering a collective approach to addressing environmental challenges associated with gene regulation technologies.

Example: Initiatives taken by international organizations aim to facilitate international cooperation in the regulation of gene editing technologies. By bringing together scientists, policymakers, and stakeholders from diverse regions, these initiatives strive to establish a common understanding and framework for the responsible use of CRISPRi in environmental applications.

To the end, as CRISPRi transforms the landscape of gene regulation, its implications for the environment necessitate a careful and multidimensional approach. The dynamic interplay between precision agriculture, bioproduction optimization, regulatory frameworks, public perception, and international collaboration defines the path forward. Striking a balance between innovation and environmental stewardship will undoubtedly shape the sustainable integration of CRISPRi in agriculture and biotechnology in the years to come.

Chapter 8: CRISPRi in Synthetic Biology

8.1 Building Synthetic Genetic Circuits

Synthetic biology, a discipline at the forefront of genetic engineering, has witnessed a paradigm shift with the advent of CRISPR interference (CRISPRi). This section explores the remarkable realm of synthetic genetic circuits, showcasing how CRISPRi has emerged as a transformative tool in sculpting these intricate molecular pathways for a myriad of applications.

Introduction to Synthetic Genetic Circuits

At the heart of synthetic biology lies the construction of artificial genetic circuits, mimicking the logic and functionality of natural biological systems. These circuits, often referred to as 'genetic devices,' are designed to perform specific tasks by arranging the expression of genes in a controlled manner. The allure of synthetic genetic circuits lies in their potential to revolutionize fields ranging from medicine to environmental management.

The CRISPRi Revolution in Circuit Design

CRISPRi has revolutionized the field of synthetic biology by providing scientists with unprecedented precision in controlling gene expression. Traditionally, constructing synthetic genetic circuits involved the integration of various genetic components, such as promoters, enhancers, and repressors. CRISPRi, however, introduces a novel layer of control by allowing researchers to selectively silence target genes.

Example 1: Toggle Switches for Cellular Memory

One fascinating application of CRISPRi in synthetic genetic circuits is the creation of toggle switches. These switches enable cells to 'remember' their state, maintaining stable gene expression patterns over time. Researchers, inspired by natural regulatory mechanisms, have harnessed CRISPRi to establish bistability within cellular populations.

In a groundbreaking study conducted by Zhang et al. (2018), a synthetic toggle switch was designed using CRISPRi in E. coli. By strategically employing CRISPRi to repress specific genes, the researchers achieved a stable ON or OFF state in the bacterial population. This toggle switch concept holds tremendous potential for developing cellular memory in engineered organisms, a crucial feature in applications like biosensors and programmable therapeutics.

Example 2: CRISPRi-Mediated Oscillators

Oscillatory behaviour is a fundamental characteristic of many biological systems, from the circadian rhythms in organisms to the cell cycle progression. Leveraging CRISPRi, researchers have successfully engineered synthetic genetic oscillators with precise control over oscillation frequencies.

In a study by Lee et al. (2019), a CRISPRi-based oscillator was constructed in yeast cells. By modulating the expression levels of key genes using CRISPRi, the researchers achieved robust and tuneable oscillations. This innovation opens avenues for developing synthetic systems with programmable temporal dynamics, holding promise for applications in biotechnology and bioengineering.

CRISPRi-Mediated Feedback Loops

Feedback loops are integral to natural biological systems, regulating processes such as homeostasis and signal transduction. CRISPRi has emerged as a powerful tool for engineering synthetic genetic circuits with precisely controlled feedback mechanisms.

Example 3: Implementing Negative Feedback Loops

Negative feedback loops play a crucial role in maintaining system stability by dampening fluctuations in gene expression. CRISPRi allows researchers to engineer synthetic circuits with precise negative feedback, ensuring fine-tuned regulation.

In a study by Wang et al. (2020), a negative feedback loop was constructed using CRISPRi in mammalian cells. By inhibiting specific genes involved in the circuit, the researchers achieved a robust negative feedback mechanism, highlighting the potential of CRISPRi in designing synthetic systems with enhanced stability and predictability.

CRISPRi-Mediated Signal Processing

Synthetic genetic circuits often require the ability to process external signals and produce specific responses. CRISPRi, with its targeted gene repression capabilities, offers a unique avenue for engineering signal processing modules within synthetic circuits.

Example 4: Signal Integration in Eukaryotic Cells

In a pioneering study by Chen et al. (2017), CRISPRi was employed to design synthetic circuits for signal integration in eukaryotic cells. By selectively inhibiting key genes in response to different signals, the researchers created a modular system capable of integrating multiple input signals and producing a specific output response. This advancement holds promise for applications in the development of responsive and programmable cellular therapies.

Future Perspectives and Challenges

While the integration of CRISPRi in synthetic genetic circuits has opened new horizons in synthetic biology, challenges and opportunities abound. Fine-tuning the CRISPRi system for optimal performance, addressing potential off-target effects, and expanding the scope of applications are areas that demand further exploration.

Final Thoughts

The marriage of CRISPRi with synthetic biology has ushered in a new era of precision and control in designing genetic circuits. From toggle switches to feedback loops and signal processing modules, CRISPRi has become an indispensable tool, empowering scientists to engineer sophisticated and responsive biological systems with profound implications for biotechnology, medicine, and beyond. As we navigate this exciting frontier, the continued synergy between CRISPRi and synthetic biology promises to unveil innovative solutions to some of the most pressing challenges in biotechnological research.

8.2 Controlling Microbial Communities

The manipulation of microbial communities using CRISPR interference (CRISPRi) has emerged as a transformative avenue within the field of synthetic biology. Microbes play pivotal roles in various ecological processes, industrial applications, and human health. Harnessing the power of CRISPRi to control and engineer microbial communities opens new possibilities for sustainable practices, bioproduction, and environmental management.

The Ecological Impact of Microbial Communities

Microbial communities, comprising diverse bacterial, archaeal, and fungal species, form intricate networks that shape ecosystems. From soil fertility to nutrient cycling, these communities are key players in maintaining ecological balance. However, harnessing their potential for human benefit requires a nuanced understanding of their dynamics.

CRISPRi as a Precision Tool for Microbial Engineering

CRISPRi offers a precise and targeted approach to manipulate gene expression in microbes. Unlike traditional genetic engineering methods, CRISPRi enables researchers to modulate the activity of specific genes without altering the underlying DNA sequence. This capability is particularly powerful in the context of microbial communities, where fine-tuning gene expression can lead to desired outcomes without disrupting the delicate balance of the ecosystem.

Engineering Microbial Communities for Bioproduction

In industrial settings, microbial communities are often harnessed for the production of biofuels, pharmaceuticals, and other valuable compounds. CRISPRi provides a revolutionary tool to optimize these communities for enhanced bioproduction. By selectively repressing or activating key genes, researchers can tailor microbial communities to increase yields, improve product quality, and streamline manufacturing processes.

Example 1: CRISPRi in Bioremediation

One striking example of CRISPRi's potential lies in the field of bioremediation. Researchers are exploring ways to use microbial communities for the cleanup of environmental pollutants. By employing CRISPRi to modulate the expression of genes involved in pollutant degradation pathways, scientists can enhance the efficiency of these microbial communities in breaking down contaminants, mitigating environmental impact.

Example 2: Enhancing Industrial Fermentation

In industrial fermentation processes, microbial communities are employed for the production of various chemicals and bio-based products. CRISPRi allows for precise control over metabolic pathways, enabling the optimization of these communities for increased yields and improved efficiency. This approach not only enhances the economic viability of industrial processes but also aligns with sustainability goals by reducing resource consumption and waste generation.

Tailoring Microbial Communities for Agriculture

In agriculture, microbial communities in the soil play crucial roles in nutrient cycling, plant health, and disease resistance. Harnessing CRISPRi to engineer these communities holds immense potential for sustainable and precision agriculture. Researchers can target specific genes to enhance nutrient uptake, confer resistance to pathogens, and improve overall soil health.

Example 3: CRISPRi for Enhanced Crop Yield

Agricultural researchers are exploring the use of CRISPRi to optimize microbial communities associated with plant roots. By modulating the expression of genes involved in nutrient mobilization and uptake, scientists aim to develop crops with increased yield and resilience. This approach not only addresses food security challenges but also minimizes the reliance on external inputs such as fertilizers.

Ethical Considerations and Environmental Impact

While the prospect of manipulating microbial communities for various applications is promising, it raises ethical considerations. Researchers and practitioners must carefully weigh the potential benefits against environmental risks. The release of engineered microbial strains into the environment could have unforeseen consequences, necessitating robust risk assessment protocols and regulatory frameworks.

Example 4: Mitigating Potential Risks in Environmental Release

In anticipating and addressing environmental risks, researchers are developing strategies to contain engineered microbial strains. CRISPRi allows for the incorporation of "kill switches" that can be activated to halt the growth and activity of engineered microbes in the event of unintended consequences. This proactive approach underscores the commitment to responsible and sustainable synthetic biology practices.

Future Directions and Challenges

As researchers continue to explore the possibilities of CRISPRi in manipulating microbial communities, several challenges and exciting avenues for future research emerge. Fine-tuning the technique for complex ecosystems, deciphering the long-term ecological impacts, and developing standardized methodologies are among the challenges. However, the potential benefits, ranging from sustainable agriculture to eco-friendly bioproduction, make these activities crucial for advancing the field of synthetic biology.

Thus, the ability of CRISPRi to manipulate microbial communities presents a paradigm shift in synthetic biology. From environmental remediation to precision agriculture, the applications are diverse and impactful. As researchers navigate the complexities of ecological systems, the responsible and ethical use of CRISPRi will be paramount in unlocking the full potential of microbial community engineering.

8.3 Bioengineering and Industrial Applications

In the fast-paced domain of bioengineering and industrial applications, the utilization of CRISPR interference (CRISPRi) has emerged as a transformative force, offering unprecedented control over gene expression for a myriad of purposes. This chapter explores the diverse ways in which CRISPRi is shaping the industrial landscape, driving innovation, and revolutionizing biotechnological processes.

Precision Engineering of Microbial Factories

At the heart of industrial biotechnology lies the quest for efficient and sustainable production methods. CRISPRi has become a linchpin in achieving precision engineering of microbial factories for biofuel production, pharmaceuticals, and other valuable compounds. By modulating gene expression, researchers can optimize metabolic pathways, enhancing the yield of desired products while minimizing unwanted byproducts. For instance, engineering strains of Escherichia coli with CRISPRi has proven instrumental in tailoring microbial metabolism for improved bioethanol production, exemplifying the potential of CRISPRi in biofuel advancements.

Accelerating Bioprocessing through CRISPRi

Bioprocessing, a critical component of industrial biotechnology, involves the use of living cells to produce valuable products on a large scale. CRISPRi's ability to precisely regulate gene expression facilitates the optimization of bioprocesses. In the production of therapeutic proteins, for example, CRISPRi enables fine-tuning of cellular machinery, ensuring high protein yields while maintaining cell viability. This not only enhances the efficiency of bioprocessing but also reduces production costs, making CRISPRi an invaluable tool in the pharmaceutical and biomanufacturing industries.

Tailoring Microorganisms for Specialty Chemicals

One of the remarkable aspects of CRISPRi in bioengineering is its versatility in tailoring microorganisms for the production of specialty chemicals. Traditional chemical synthesis methods often involve environmentally harmful processes and yield undesirable byproducts. CRISPRi offers a sustainable alternative by allowing researchers to reprogram microbial hosts for the synthesis of specific chemicals. For instance, manipulating gene expression in yeast through CRISPRi has been employed to create strains capable of producing bio-based chemicals, presenting a greener and more eco-friendly approach to chemical manufacturing.

Optimizing Agricultural Biotechnology with CRISPRi

In the realm of agricultural biotechnology, CRISPRi holds the potential to revolutionize crop improvement and protection. By precisely modulating the expression of genes related to stress resistance, crop yield, and nutritional content, CRISPRi enables the creation of genetically modified organisms (GMOs) with enhanced traits. This not only contributes to global food security but also addresses challenges such as climate change and resource limitations. The application of CRISPRi in agriculture underscores its role in shaping a more sustainable and resilient agricultural sector.

Redefining Synthetic Biology in Industry

Synthetic biology, the amalgamation of biology and engineering principles to design and construct new biological entities, has found a powerful ally in CRISPRi. The precise control over gene expression provided by CRISPRi facilitates the construction of synthetic genetic circuits with intricate functionalities. This capability has far-reaching implications in the production of bio-based materials, chemicals, and even therapeutic proteins. Researchers can now engineer microbial hosts to execute complex tasks, such as sensing environmental cues and responding with specific metabolic activities, ushering in a new era of programmable biology for industrial applications.

Addressing Challenges and Future Prospects

While the application of CRISPRi in bioengineering and industrial settings holds great promise, challenges persist. Off-target effects, delivery methods, and scalability issues are among the hurdles that researchers must address to fully unlock the potential of CRISPRi in industry. Moreover, navigating regulatory landscapes and ensuring public acceptance are crucial considerations as the technology advances.

Looking forward, the future of CRISPRi in bioengineering and industrial applications is bright. Continued research efforts are focused on refining the technology, expanding its applicability, and addressing current limitations. Collaborations between academia and industry play a pivotal role in translating laboratory discoveries into real-world applications, ensuring that CRISPRi continues to shape the landscape of bioengineering, providing sustainable solutions to pressing global challenges. As CRISPRi evolves, its impact on industrial biotechnology promises to be both profound and enduring.

Chapter 9: CRISPRi in Neurobiology

9.1 Gene Regulation in the Brain

The potential of gene regulation within the brain arranges the opus of neural development, function, and response to environmental stimuli. The complexity of the human brain, with its billions of interconnected neurons, relies on precise gene expression to shape cognition, behaviour, and the foundation of neurological health. This section delves into the profound impact of CRISPR interference (CRISPRi) in unravelling the mysteries of gene regulation within the enigmatic world of neurobiology.

Unravelling the Genetic Tapestry of Neurological Function

The brain, a marvel of biological engineering, exhibits a highly regulated gene expression landscape crucial for its diverse functions. Neurodevelopment, synaptic plasticity, and response to external stimuli are modulated by an intricate interplay of genes. CRISPRi emerges as a powerful tool to decipher this complexity by offering precise control over gene expression levels.

Precision in Targeting Neurological Genes

One of the paramount challenges in neurobiology lies in identifying and manipulating specific genes implicated in neural function. CRISPRi, armed with its guide RNA (gRNA) precision, enables researchers to target and inhibit the expression of genes with unprecedented accuracy. For instance, a groundbreaking study led by Li et al. (2019) successfully employed CRISPRi to selectively downregulate the expression of the BDNF gene, a key player in synaptic plasticity. This precise targeting illuminated the gene's role in shaping neural networks and underscored the potential of CRISPRi in dissecting intricate neurobiological pathways.

Unmasking the Secrets of Neurological Disorders

Neurological disorders, ranging from Alzheimer's to schizophrenia, often have a genetic basis. CRISPRi has emerged as a transformative tool in modelling these disorders by allowing researchers to manipulate gene expression and observe the ensuing cellular and molecular consequences. In a landmark study by Wang et al. (2021), CRISPRi was utilized to modulate the expression of genes associated with autism spectrum disorders. The results provided valuable insights into the underlying mechanisms, paving the way for potential therapeutic interventions.

CRISPRi's Therapeutic Potential for Brain Diseases

Beyond its role in understanding the molecular nuances of the brain, CRISPRi holds immense promise in the therapeutic realm. Neurological diseases, often characterized by dysregulated gene expression, present a compelling target for CRISPRi-based interventions.

Navigating the Blood-Brain Barrier

The blood-brain barrier (BBB) poses a formidable challenge in delivering therapeutic agents to the brain. CRISPRi, with its non-viral delivery systems and localized gene regulation capabilities, offers a promising avenue to overcome this hurdle. Recent advancements, such as the use of adeno-associated viruses (AAVs) for CRISPRi delivery, showcase the potential for targeted and efficient gene regulation within the brain's intricate confines.

Addressing Genetic Culprits of Neurodegeneration

Neurodegenerative diseases, characterized by the progressive loss of neuronal function, often have a genetic basis. CRISPRi presents an opportunity to intervene at the genetic level, potentially slowing or halting the progression of these debilitating conditions. A study by Zhang et al. (2020) demonstrated the feasibility of using CRISPRi to downregulate genes associated with Parkinson's disease, offering a glimpse into a future where gene regulation therapies could mitigate the impact of neurodegenerative disorders.

Charting the Course for Future Discoveries

The journey into understanding gene regulation in the brain through CRISPRi is just beginning, promising a landscape of discoveries that could redefine our approach to neurological health.

Epigenetics and Beyond

As the interplay between genetics and epigenetics becomes increasingly apparent in neurological function, CRISPRi stands poised to play a pivotal role in deciphering these complex interactions. By selectively modifying gene expression, CRISPRi allows researchers to explore the dynamic nature of epigenetic marks and their influence on neural development and function.

Beyond Single Genes: Systems Neurobiology

The integration of CRISPRi with systems biology approaches holds the potential to unravel the intricate networks governing brain function. By simultaneously regulating multiple genes, researchers can gain a holistic understanding of how gene networks orchestrate neural processes. This shift from a reductionist to a systems-level approach opens new vistas for comprehending the interconnected web of genes that define the brain's functionality.

Thus, CRISPRi emerges as a beacon of precision and promise in the quest to understand and modulate gene regulation in the brain. From untying the details of neural development to offering therapeutic avenues for neurological diseases, CRISPRi stands at the forefront of neurobiological research, providing a roadmap for the future understanding and manipulation of the brain's genetic home.

9.2 Investigating Neurological Disorders

The exploration of CRISPR interference (CRISPRi) has ignited a transformative journey in the quest of understanding and addressing complex challenges within the complex aspects of neurological disorders. This section digs into the application of CRISPRi in separating the particulars of neurological disorders, shedding light on its potential for elucidating the underlying genetic factors, exploring therapeutic avenues, and reshaping the landscape of neurobiology research.

Neurological disorders, encompassing a spectrum from neurodevelopmental conditions to neurodegenerative diseases, pose a significant burden on global health. The genetic underpinnings of these disorders have long been a subject of intense investigation. CRISPRi emerges as a powerful tool in this pursuit, allowing researchers to meticulously decipher the genetic basis of neurological disorders with unprecedented precision.

Targeting Disease-Associated Genes

One of the primary applications of CRISPRi in neurobiology lies in the targeted regulation of genes implicated in neurological disorders. By utilizing a catalytically dead Cas9 (dCas9) protein coupled with guide RNAs (gRNAs) designed to bind specific genomic loci, researchers can modulate the expression of genes associated with conditions such as Alzheimer's disease, Parkinson's disease, and autism spectrum disorders.

For instance, studies have employed CRISPRi to investigate the role of the APOE gene in Alzheimer's disease. The APOE gene encodes a protein involved in lipid transport, and certain variants of this gene are known to increase the risk of developing Alzheimer's. Using CRISPRi, researchers have selectively inhibited the expression of APOE, observing changes in cellular phenotypes and shedding light on the molecular pathways influenced by this gene.

Functional Genomics and Pathway Analysis

Beyond individual gene modulation, CRISPRi facilitates large-scale functional genomics studies to elucidate the broader molecular landscape associated with neurological disorders. By systematically inhibiting genes across the genome and observing resulting phenotypic changes, researchers gain insights into the intricate networks and pathways contributing to disease pathology.

In a noteworthy study focusing on Parkinson's disease, CRISPRi was employed to systematically target genes in dopaminergic neurons. This approach unveiled novel genes involved in neuronal survival and function, providing a comprehensive view of the genetic factors influencing the susceptibility and progression of Parkinson's disease. Such high-throughput analyses contribute to building a comprehensive map of the genetic circuitry underlying neurological disorders.

Recapitulating Disease Phenotypes in vitro

CRISPRi extends its impact beyond the analysis of endogenous gene expression by enabling the creation of in vitro models that recapitulate disease phenotypes. Neurological disorders often present challenges in studying disease mechanisms directly in human brains. CRISPRi allows researchers to introduce disease-associated genetic alterations into human-derived cell lines or induced pluripotent stem cells, providing a platform for detailed mechanistic studies.

For instance, in the context of amyotrophic lateral sclerosis (ALS), CRISPRi has been utilized to modulate the expression of genes linked to the disease in motor neurons derived from induced pluripotent stem cells. This approach not only replicated key features of ALS pathology but also allowed for the screening of potential therapeutic compounds, showcasing the translational potential of CRISPRi in drug discovery for neurological disorders.

Unscrambling the Complexity of Polygenic Disorders

Many neurological disorders, such as schizophrenia and epilepsy, are polygenic, involving the interplay of multiple genetic factors. CRISPRi facilitates the dissection of this complexity by enabling the simultaneous regulation of multiple genes associated with these conditions.

In a groundbreaking study focused on schizophrenia, researchers utilized CRISPRi to target a set of risk genes identified through genome-wide association studies. By modulating the expression of these genes individually and in combination, researchers were able to uncover synergistic effects and interactions that contribute to the manifestation of schizophrenia-related phenotypes. This approach highlights the potential of CRISPRi in deciphering the intricate genetic architecture of polygenic neurological disorders.

Towards Therapeutic Intervention

Beyond its role in uncovering the genetic basis of neurological disorders, CRISPRi holds promise as a tool for developing therapeutic interventions. The ability to selectively and reversibly modulate gene expression provides a unique avenue for exploring novel treatment strategies.

Precision Medicine Approaches

CRISPRi opens the door to precision medicine approaches in the treatment of neurological disorders. By identifying and validating specific gene targets associated with an individual's disease profile, researchers can envision developing personalized therapeutic strategies. For instance, in the case of epilepsy, CRISPRi has been employed to selectively modulate ion channels and receptors implicated in seizure activity, paving the way for precision therapies tailored to the underlying genetic factors in individual patients.

CRISPRi in Gene Therapy

The potential of CRISPRi in gene therapy for neurological disorders is an area of active exploration. Unlike traditional gene editing approaches that permanently modify the genome, CRISPRi offers a reversible and fine-tuneable method for modulating gene expression. This characteristic is particularly advantageous in conditions where tight control of gene expression is essential.

In preclinical studies related to Huntington's disease, CRISPRi has been employed to target the mutant huntingtin gene responsible for the disease. By suppressing the expression of the mutant gene, researchers observed a reduction in disease-associated phenotypes in animal models. This approach exemplifies the therapeutic potential of CRISPRi in mitigating the impact of disease-causing genes without permanently altering the genomic landscape.

Final Thoughts

The exploration of CRISPRi in investigating neurological disorders marks a paradigm shift in the landscape of neurobiology research. From deciphering the intricate genetic basis of these disorders to envisioning novel therapeutic interventions, CRISPRi has emerged as a versatile and powerful tool. As research continues to advance, the integration of CRISPRi with other cutting-edge technologies promises a more comprehensive understanding of the molecular dance underlying neurological disorders, propelling us closer to effective treatments and personalized solutions for individuals facing these challenging conditions.

9.3 Therapeutic Potential for Brain Diseases

In recent years, the application of CRISPR interference (CRISPRi) in neurobiology has sparked considerable excitement, particularly in the realm of therapeutic interventions for brain diseases. The intricate nature of neurological disorders, ranging from Alzheimer's disease to various forms of epilepsy, presents a formidable challenge for traditional treatment approaches. Here, we explore how CRISPRi, with its precision gene-regulating capabilities, holds promise for transforming the landscape of therapeutic strategies targeting the intricate machinery of the brain.

Precision Gene Regulation in the Brain

The human brain, with its billions of interconnected neurons, orchestrates the symphony of thoughts, emotions, and bodily functions. Neurological disorders often manifest when this symphony is disrupted, leading to a cascade of molecular events. CRISPRi offers a unique advantage in this context, allowing researchers to selectively modulate the expression of specific genes implicated in various brain diseases.

One promising avenue is the application of CRISPRi in targeting genes associated with neurodegenerative diseases. For instance, the aberrant accumulation of amyloid-beta peptides is a hallmark of Alzheimer's disease. By employing CRISPRi, researchers have successfully downregulated genes responsible for the production of these peptides, mitigating pathological changes in experimental models. This precision gene regulation opens new possibilities for developing therapies that address the root causes of neurodegenerative disorders.

Investigating Neurological Disorders at the Genetic Level

CRISPRi not only provides a means to modulate gene expression but also serves as a powerful tool for unravelling the genetic minutiae underlying neurological disorders. Through the systematic inhibition of specific genes associated with conditions such as Parkinson's disease and Huntington's disease, researchers gain insights into the molecular pathways driving these disorders.

In a groundbreaking study, scientists used CRISPRi to silence genes associated with Parkinson's disease in neurons derived from patient-specific induced pluripotent stem cells. By observing the resulting changes in cellular function, researchers identified novel targets for therapeutic intervention. This approach not only deepens our understanding of the genetic basis of neurological disorders but also facilitates the development of targeted treatments tailored to individual patients.

CRISPRi as a Therapeutic Tool for Brain Diseases

The therapeutic potential of CRISPRi extends beyond elucidating the genetic underpinnings of neurological disorders—it offers a novel and precise avenue for treatment. Various studies have demonstrated the feasibility of using CRISPRi to modulate gene expression in vivo, opening the door to potential therapeutic applications.

In the context of epilepsy, a disorder characterized by abnormal electrical activity in the brain, CRISPRi has shown promise in mitigating seizures. By selectively inhibiting genes associated with hyperexcitability in neurons, researchers have successfully reduced seizure frequency in animal models. This targeted approach holds immense therapeutic potential, as it addresses the specific molecular mechanisms contributing to epilepsy.

Furthermore, CRISPRi has shown efficacy in addressing genetic mutations associated with rare neurodevelopmental disorders. For instance, Rett syndrome, caused by mutations in the MECP2 gene, presents challenges for traditional therapeutic approaches. CRISPRi allows for the selective silencing of the mutated MECP2 gene, offering a potential avenue for alleviating symptoms associated with the disorder.

Challenges and Future Directions

While the therapeutic potential of CRISPRi in treating brain diseases is promising, challenges and ethical considerations remain. Off-target effects, albeit minimized with advanced CRISPRi techniques, necessitate continued refinement to ensure precision in gene regulation. Additionally, the delivery of CRISPR components to specific regions of the brain poses a technical challenge that requires innovative solutions.

Moreover, the ethical implications of gene editing in the brain demand careful consideration. Ensuring the safety and efficacy of CRISPRi interventions in humans requires comprehensive preclinical studies and adherence to rigorous ethical standards. Striking a balance between advancing therapeutic innovations and safeguarding against unintended consequences is paramount.

To sum up, CRISPRi stands at the forefront of a transformative era in neurobiology, offering unprecedented opportunities for investigating and treating brain diseases. The precision, versatility, and therapeutic potential of CRISPRi provide a compelling narrative for its continued exploration in the pursuit of effective treatments for complex neurological disorders. As we navigate this exciting frontier, the integration of cutting-edge science with ethical considerations will shape the future landscape of CRISPRi-based therapies for the benefit of individuals affected by brain diseases.

Chapter 10: CRISPRi in Cancer Research

10.1 Targeting Oncogenes with CRISPRi

In the age of advanced molecular biology, the role of oncogenes stands out as a key player in the genesis and progression of cancer. These genes, often mutated or overexpressed, drive the uncontrolled cellular proliferation characteristic of malignant tumours. Harnessing the precision of CRISPRi technology in the pursuit of oncogene regulation has opened new avenues in cancer research and therapeutic development.

Oncogenes, such as MYC, KRAS, and TP53, have been focal points in the exploration of CRISPRi's potential. The MYC oncogene, for instance, is notoriously implicated in a variety of cancers, including breast, colon, and lung cancers. Traditional approaches to target these genes often face challenges due to the broad and vital roles these genes play in normal cellular function. CRISPRi, with its ability to specifically modulate gene expression without altering the underlying DNA sequence, provides a nuanced tool for scientists to navigate the intricate genetic landscape of cancer.

One notable success in the application of CRISPRi to oncogene regulation is the modulation of the MYC gene. MYC, a transcription factor, regulates the expression of numerous downstream genes involved in cell growth and division. Its overexpression is a common feature in many cancer types. Researchers have employed CRISPRi to precisely dial down the expression of MYC, resulting in a slowdown of cancer cell proliferation. This approach showcases the potential of CRISPRi in fine-tuning gene expression to achieve therapeutic outcomes.

In a groundbreaking study published in *Nature Communications*, researchers used CRISPRi to target the MYC gene in pancreatic cancer cells. By designing specific guide RNAs to bind to the promoter region of MYC, they effectively silenced its expression without inducing permanent genetic changes. This reversible and targeted regulation demonstrated a remarkable reduction in the growth of pancreatic cancer cells in preclinical models.

Similarly, the notorious KRAS oncogene, frequently mutated in various cancers, has been a challenging target for traditional therapeutic strategies. CRISPRi offers a promising avenue to selectively inhibit KRAS expression, hindering the signalling cascades that drive uncontrolled cell division. In a study published in *Science Advances*, scientists utilized CRISPRi to downregulate KRAS in colorectal cancer cells, leading to a significant decrease in tumour growth.

The versatility of CRISPRi is not confined to a singular gene but extends to multifaceted networks involved in cancer progression. TP53, a tumour suppressor gene commonly mutated in cancers, serves as another compelling example. Traditional gene-editing approaches may inadvertently disrupt normal TP53 function. In contrast, CRISPRi allows for the nuanced control of TP53 expression, enabling researchers to explore therapeutic interventions without compromising essential cellular functions.

The application of CRISPRi in oncogene regulation not only holds promise for basic research but also paves the way for innovative cancer therapies. The dynamic landscape of cancer genetics, characterized by genetic heterogeneity even within a single tumour, necessitates precision in therapeutic interventions. CRISPRi's ability to modulate gene expression at the transcriptional level provides a strategic advantage in addressing this complexity.

Beyond its potential in the laboratory, CRISPRi-based cancer therapies are progressing towards clinical trials. The prospect of selectively targeting oncogenes with minimal off-target effects presents a paradigm shift in cancer treatment. CRISPRi-based therapies may offer a personalized approach, tailoring interventions to the unique genetic makeup of each patient's cancer.

However, challenges persist on the road to clinical translation. Off-target effects, though minimized with CRISPRi compared to traditional CRISPR-Cas9 approaches, remain a consideration. Ensuring the safety and efficacy of CRISPRi-based therapies requires meticulous optimization of guide RNA designs and delivery methods. These challenges underscore the importance of continued research to refine and enhance the precision of CRISPRi technology in the context of cancer therapeutics.

Therefore, the targeted modulation of oncogenes with CRISPRi represents a transformative chapter in the ever-evolving field of cancer research. From MYC to KRAS and TP53, CRISPRi's precision in gene regulation has illuminated new possibilities for understanding and treating cancer. The journey from laboratory discoveries to clinical applications is ongoing, holding the promise of a future where the symphony of molecular biology is orchestrated to bring harmony to the complex world of cancer.

10.2 Gene Regulation in Tumour Suppression

Cancer, a complex and relentless adversary, continues to challenge the frontiers of medical science. Within this formidable terrain, the exploration of CRISPR interference (CRISPRi) emerges as a promising ally in the pursuit of innovative cancer therapies. One of the pivotal domains where CRISPRi exhibits significant potential is in the regulation of genes implicated in tumour suppression.

Understanding Tumour Suppressor Genes

Tumour suppressor genes, the guardians of genomic integrity, play a pivotal role in preventing uncontrolled cell proliferation and the onset of malignancy. Examples abound, with the iconic TP53 and BRCA1 genes standing as sentinels against the unchecked growth that characterizes cancer. These genes encode proteins crucial for maintaining cellular homeostasis, orchestrating DNA repair, and orchestrating programmed cell death, or apoptosis.

CRISPRi as a Precision Scalpel

CRISPRi's precision in targeting specific genomic loci positions it as a molecular scalpel for modulating gene expression. In the context of tumour suppression, this means the ability to selectively inhibit or fine-tune the activity of specific genes implicated in cancer development. For instance, a well-crafted guide RNA (gRNA) designed to target the promoter region of a tumour suppressor gene can lead to controlled transcriptional repression without permanently altering the underlying genetic code.

Targeting TP53: A Paradigm in Tumour Suppression

The TP53 gene, often referred to as the "guardian of the genome," exemplifies the potential impact of CRISPRi in cancer research. By harnessing CRISPRi, researchers can precisely modulate TP53 expression levels, mimicking the natural regulatory mechanisms that govern its function. This not only deepens our understanding of TP53's role but also opens avenues for therapeutic interventions.

Navigating the BRCA1 Landscape

In the landscape of breast and ovarian cancers, the BRCA1 gene assumes a central role. Mutations in BRCA1 predispose individuals to a higher risk of developing these cancers. CRISPRi facilitates nuanced exploration, allowing researchers to manipulate BRCA1 expression levels in a controlled manner. This offers insights into the gene's intricate regulatory network and its implications for tumour suppression.

Fine-Tuning Apoptosis with CRISPRi

Apoptosis, programmed cell death, serves as a fundamental defence mechanism against tumorigenesis. Key regulators of apoptosis, such as the BCL2 family, are prime candidates for CRISPRi-based investigations. By selectively modulating the expression of anti-apoptotic genes like BCL2, researchers can uncover novel strategies to induce apoptosis in cancer cells, potentially providing a therapeutic avenue.

CRISPRi Screens: Mining for Tumour Suppressor Gold

Large-scale CRISPRi screens have become instrumental in identifying novel players in tumour suppression. These screens involve systematically inhibiting the expression of various genes and observing the impact on cellular proliferation and survival. Through such activities, researchers have unearthed genes that, when downregulated, exhibit tumour-suppressive properties, expanding the repertoire of potential therapeutic targets.

Overcoming Resistance: The Achilles Heel of Tumours

Resistance to traditional cancer therapies often arises due to the adaptive nature of tumours. CRISPRi offers a means to decipher the genetic underpinnings of resistance mechanisms. By selectively targeting genes associated with resistance, such as those involved in drug efflux or DNA repair pathways, CRISPRi contributes valuable insights to enhance the efficacy of existing treatments.

Epigenetic Modulation: Unveiling Hidden Tumour Suppressors

Beyond the traditional focus on genetic alterations, CRISPRi's influence extends to the realm of epigenetics. Aberrant DNA methylation and histone modifications can silence tumour suppressor genes. CRISPRi allows for the targeted removal of these epigenetic marks, reactivating silenced tumour suppressors and providing a novel avenue for therapeutic exploration.

From Bench to Bedside: Clinical Implications

The transition from laboratory discovery to clinical application is a critical juncture in any therapeutic pursuit. CRISPRi's potential in tumour suppression is no exception. Clinical trials exploring the feasibility and safety of CRISPRi-based therapies for cancer are underway. The hope is that these endeavors will pave the way for a new era in precision oncology, where the unique genetic signature of each patient's tumour informs targeted therapeutic interventions.

Ethical Considerations in Gene Regulation for Cancer

As we navigate the exciting possibilities offered by CRISPRi in cancer research, ethical considerations loom large. Questions surrounding the responsible use of gene-editing technologies, potential off-target effects, and the implications of heritable changes demand careful scrutiny. Striking a balance between scientific advancement and ethical responsibility remains imperative as the field progresses.

Final Thoughts

The exploration of CRISPRi in the realm of tumour suppression is an unfolding narrative of scientific discovery and potential therapeutic breakthroughs. From untying the details of well-known tumour suppressors to identifying novel candidates, CRISPRi's impact reverberates across the landscape of cancer research. As we stand on the cusp of a new era in gene regulation, the journey into the genetic arena of tumour suppression with CRISPRi promises to illuminate novel pathways toward effective cancer treatments.

10.3 Future Avenues for Cancer Treatment

The intersection of CRISPR interference (CRISPRi) and cancer research has illuminated promising avenues for innovative therapeutic strategies. As we traverse the complex scenery of gene regulation, the role of CRISPRi in shaping the future of cancer treatment becomes increasingly apparent. This section investigates into the dynamic developments and future prospects dignified to revolutionize oncological interventions.

Targeting Elusive Oncogenes

One of the paramount challenges in cancer treatment has been the identification and targeting of elusive oncogenes responsible for tumorigenesis. Traditional therapeutic approaches often fall short in selectively inhibiting these oncogenic culprits. CRISPRi, with its precision gene-silencing capabilities, offers a tailored solution. Recent studies have demonstrated the successful inhibition of specific oncogenes using CRISPRi, providing a beacon of hope for patients with cancers driven by hard-to-reach genetic mutations.

For instance, the MYC oncogene, notorious for its role in various cancers, has long eluded targeted therapies. CRISPRi has emerged as a potent tool to modulate MYC expression, hindering tumour growth in preclinical models. This breakthrough not only showcases the therapeutic potential of CRISPRi but also hints at a paradigm shift in how we approach and treat malignancies driven by elusive oncogenes.

Overcoming Resistance to Conventional Therapies

The emergence of resistance to conventional cancer therapies poses a formidable obstacle in the clinical management of the disease. As cancer cells adapt and evolve, they often develop mechanisms to circumvent the effects of chemotherapy and targeted therapies. CRISPRi provides a multifaceted strategy to overcome resistance by disrupting the genetic pathways that confer resistance to standard treatments.

A poignant example lies in the context of targeted therapies against tyrosine kinase receptors in certain cancers. While these therapies initially exhibit efficacy, resistance inevitably emerges. CRISPRi allows researchers to pinpoint and silence the genes responsible for this acquired resistance, restoring the sensitivity of cancer cells to targeted interventions. This approach not only extends the utility of existing therapies but also introduces a level of adaptability crucial for navigating the evolving landscape of cancer biology.

Personalized Therapies Tailored to Genomic Signatures

The era of precision medicine is on the horizon, and CRISPRi stands at the forefront of tailoring therapies to individual genomic signatures. The heterogeneity of cancer, even within a single type, underscores the need for personalized treatment strategies. CRISPRi enables researchers to conduct comprehensive genomic profiling and design therapies that specifically target the unique genetic alterations driving an individual's cancer.

In a groundbreaking study, researchers utilized CRISPRi to modulate the expression of multiple genes simultaneously, creating a bespoke therapeutic cocktail for each patient. This approach, termed "genomic surgery," holds immense promise for personalized cancer treatment. By exploiting CRISPRi's capacity for multiplex gene regulation, clinicians may soon administer precisely tailored therapies that address the specific genetic landscape of each patient's tumour.

Unleashing the Potential of Immunotherapy

The advent of immunotherapy has revolutionized cancer treatment by harnessing the body's immune system to target and eliminate cancer cells. However, not all patients respond favourably to immunotherapy, and enhancing its efficacy remains a pressing challenge. CRISPRi emerges as a powerful ally in this effort by fine-tuning the expression of genes involved in the immune response.

Recent studies have employed CRISPRi to modulate immune checkpoint genes, such as PD-1 and CTLA-4, with remarkable success. By downregulating these inhibitory signals, CRISPRi potentiates the immune system's ability to recognize and eradicate cancer cells. The synergy between CRISPRi and immunotherapy represents a burgeoning field of exploration, promising to elevate the effectiveness of immunomodulatory approaches and broaden the spectrum of patients benefiting from these therapies.

Navigating the Regulatory Landscape

As CRISPRi transitions from the laboratory bench to potential clinical applications, navigating the regulatory landscape becomes a pivotal consideration. Regulatory bodies worldwide are grappling with the ethical and safety implications of gene-editing technologies in therapeutic contexts. Striking a balance between promoting innovation and ensuring patient safety remains a paramount challenge.

The prospect of utilizing CRISPRi in cancer treatment necessitates collaborative efforts among researchers, clinicians, ethicists, and regulatory agencies. Establishing transparent frameworks for the ethical and responsible use of CRISPRi in clinical trials and eventual therapeutic applications is imperative. Concerted initiatives to address these ethical considerations will be instrumental in fostering public trust and facilitating the seamless integration of CRISPRi into the arsenal of cancer treatment modalities.

A Transformative Frontier in Cancer Treatment

The future of cancer treatment is intricately intertwined with the continued exploration of CRISPR interference. From targeting elusive oncogenes to overcoming resistance and paving the way for personalized therapies, CRISPRi emerges as a transformative force in the fight against cancer. As research advances and clinical trials progress, the potential for CRISPRi to redefine the landscape of oncology becomes increasingly evident. The dynamic interplay between scientific innovation, clinical implementation, and ethical considerations will shape the trajectory of CRISPRi as it evolves into a cornerstone of next-generation cancer therapeutics.

Chapter 11: Regulatory Mechanisms of CRISPRi

11.1 Understanding CRISPRi Off-Switches

In the jumble of gene regulation, the quest for precision and control has led scientists to unravel the subtleties of CRISPR interference's (CRISPRi) off-switches. As we circumnavigate this complex genetic terrain, understanding the mechanisms that dictate when to halt CRISPRi becomes crucial for fine-tuning gene expression and mitigating unintended consequences.

The Intricacies of CRISPRi Off-Switches: A Molecular Ballet

To comprehend CRISPRi off-switches, we must first recognize the orchestrated molecular ballet within the cell. At the heart of CRISPRi lies the deactivated Cas9 (dCas9), a protein scaffold that acts as a conductor orchestrating gene regulation. The dCas9's primary accomplice is the guide RNA (gRNA), a molecular partner that guides dCas9 to specific genomic locations.

The off-switches in CRISPRi are essentially the strategies employed to halt this molecular ballet. One notable off-switch mechanism involves the reversible binding of dCas9 to its target DNA. This dance of binding and unbinding is tightly regulated, allowing researchers to modulate gene expression with exquisite precision.

Active Relationship: The Role of Epigenetic Modifications

While taking into consideration the landscape of gene regulation, epigenetic modifications emerge as key players influencing CRISPRi off-switches. DNA methylation and histone modifications, which dynamically shape the chromatin architecture, dictate the accessibility of the target DNA.

Studies have revealed that the interplay between CRISPRi and epigenetic modifications is not a linear narrative but rather a dynamic dialogue. For instance, certain off-switches operate more effectively in regions of open chromatin, while others might struggle to navigate condensed chromatin structures. Understanding this dynamic interplay is essential for anticipating the responsiveness of CRISPRi off-switches in different genomic contexts.

Temperature-Controlled Off-Switches: An Intriguing Twist

In the chase of expanding CRISPRi's versatility, researchers have unearthed an intriguing off-switch strategy involving temperature control. Leveraging temperature-sensitive dCas9 variants, scientists can manipulate the on/off state of CRISPRi with external temperature cues. This innovation allows for precise temporal control, opening new avenues for dynamic gene regulation in diverse experimental settings.

RNA-Mediated Off-Switches: Silence in Symphony

Adding another layer to the complexity of CRISPRi off-switches, RNA-mediated mechanisms have emerged as silent conductors orchestrating gene silencing. Through the incorporation of additional RNA elements, researchers can design off-switches that respond to endogenous RNA signals, leading to a nuanced and context-dependent regulation of gene expression.

Addressing Off-Target Effects: A Balancing Act

As we steer the landscape of CRISPRi off-switches, the looming concern of off-target effects demands our attention. Researchers are dedicatedly developing strategies to minimize unintended consequences, ensuring that the molecular ballet of CRISPRi remains focused on the intended genomic targets. Techniques such as rational gRNA design and thorough off-target prediction algorithms contribute to this ongoing effort.

Clinical Implications: Fine-Tuning for Therapeutic Precision

Understanding CRISPRi off-switches is not merely an academic pursuit but holds profound implications for therapeutic applications. In the realm of gene therapy, where precision is paramount, the ability to precisely control CRISPRi becomes a linchpin. Harnessing the knowledge of off-switch mechanisms allows researchers to fine-tune gene expression with surgical precision, minimizing the risk of unintended consequences in therapeutic interventions.

Despite significant strides in deciphering CRISPRi off-switches, challenges persist. The nuanced nature of gene regulation demands a continual exploration of off-switch mechanisms across diverse genetic contexts. Additionally, the translation of these findings into practical applications, especially in therapeutic settings, necessitates overcoming existing technical and ethical hurdles.

To sum up, this exploration into CRISPRi off-switches, it is evident that we stand on the precipice of a new era in gene regulation. The molecular ballet within the cell, guided by the delicate interplay of dCas9 and gRNA, is a captivating symphony waiting to be further understood and harnessed. In the ever-evolving landscape of CRISPR technology, the quest for precision continues, opening doors to novel therapeutic interventions and pushing the boundaries of our understanding of the genetic dance that shapes life itself.

11.2 Fine-Tuning Gene Expression

Potentializing the significance of cellular processes, precision is paramount. CRISPR interference (CRISPRi) serves as a molecular maestro, allowing scientists to delicately fine-tune gene expression with a level of control that was previously unimaginable. This section explores the nuances of fine-tuning gene expression using CRISPRi, shedding light on the practical applications, challenges, and promising outcomes.

Understanding the Dynamics of Fine-Tuning

Fine-tuning gene expression is akin to adjusting the volume on a sophisticated stereo system, where precision is essential to achieve the desired outcome. CRISPRi, with its ability to modulate gene expression with precision, provides researchers with a powerful tool for this purpose. Unlike traditional knockout methods that completely silence a gene, CRISPRi allows for the nuanced adjustment of gene expression levels, creating a spectrum of effects from subtle dampening to more pronounced inhibition.

Applications in Disease Modelling and Drug Discovery

One of the most compelling applications of fine-tuning gene expression lies in disease modelling and drug discovery. Diseases often result from dysregulated gene expression, and CRISPRi offers a way to mimic these subtle variations in the laboratory setting. For instance, in cancer research, scientists can use CRISPRi to selectively downregulate oncogenes, mirroring the intricate interplay of gene expression seen in diseased tissues. This approach facilitates the development of more accurate disease models, paving the way for targeted drug discovery and personalized medicine.

The CRISPRi Toolbox for Precision Editing

Fine-tuning gene expression demands a sophisticated toolbox, and CRISPRi provides an array of tools to achieve this. The use of catalytically inactive Cas9 (dCas9) as the "molecular scissors" allows researchers to precisely position the CRISPRi complex at specific locations within the genome. Coupled with the design of guide RNAs (gRNAs) tailored to target specific genes, this toolbox empowers scientists to delicately adjust gene expression levels.

Case Study: Neurobiology and the Art of Fine-Tuning

In neurobiology, where precise gene expression is crucial for the development and function of the nervous system, CRISPRi has emerged as a transformative tool. Researchers can finely modulate the expression of genes involved in neural development, synaptic plasticity, and neurotransmitter regulation. This level of control opens new avenues for understanding complex neurological disorders such as Alzheimer's or Parkinson's disease, where subtle variations in gene expression may contribute to disease progression.

Fine-Tuning for Therapeutic Efficacy

The therapeutic potential of CRISPRi extends beyond disease modelling to the realm of treatment. Fine-tuning gene expression becomes a therapeutic dance, where the goal is not only to silence a problematic gene but to achieve an optimal level of expression for therapeutic efficacy. For example, in gene therapy, where the introduction of a functional gene is often required, CRISPRi can delicately modulate the expression of the introduced gene to avoid potential side effects associated with overexpression.

Challenges in Precision: Off-Target Effects and Specificity

While CRISPRi offers unprecedented precision, challenges persist in achieving absolute specificity. Off-target effects, where the CRISPRi complex inadvertently interferes with unintended genes, remain a concern. Researchers are continually refining the design of gRNAs and optimizing delivery methods to minimize off-target effects, ensuring that the fine-tuning achieved is as specific as possible.

Quantifying Precision: Tools for Assessing Gene Expression Levels

Measuring the success of fine-tuning efforts requires sophisticated tools for quantifying gene expression levels. Techniques such as RNA sequencing and quantitative polymerase chain reaction (qPCR) provide researchers with the ability to precisely assess changes in gene expression. Integrating these tools into CRISPRi experiments allows for a quantitative evaluation of the extent to which gene expression has been fine-tuned.

Ethical Considerations in Gene Expression Manipulation

The power to finely modulate gene expression raises ethical questions regarding the extent to which humans should intervene in the natural order of genetic regulation. While the potential for therapeutic breakthroughs is vast, the responsible use of CRISPRi demands careful consideration of the ethical implications. Striking a balance between scientific advancement and ethical responsibility becomes paramount as researchers venture into the realm of fine-tuning the very building blocks of life.

Educating Stakeholders: Bridging the Gap Between Science and Society

As fine-tuning gene expression becomes a tangible reality, the need for educating stakeholders, including the general public, policymakers, and ethicists, becomes imperative. Transparent communication about the capabilities and limitations of CRISPRi in gene regulation ensures that societal discussions are informed and decisions regarding its application are made collectively.

Future Directions: Pushing the Boundaries of Precision

The journey of fine-tuning gene expression with CRISPRi is an ever-evolving narrative. Future research directions aim to push the boundaries of precision even further. Innovations in gRNA design, delivery methods, and the development of next-generation CRISPR tools promise to enhance the fine-tuning capabilities of CRISPRi. This relentless pursuit of precision opens new vistas for understanding the intricacies of gene regulation and holds the potential to revolutionize fields ranging from medicine to agriculture.

In the massive arena of molecular biology, fine-tuning gene expression with CRISPRi emerges as a remarkable brushstroke. This chapter has unveiled the art and science behind this precision, showcasing how researchers delicately manipulate the intricate dance of genes. As CRISPRi continues to shape the landscape of genetic research, the ability to finely modulate gene expression stands as a testament to the transformative potential of this groundbreaking technology. The coming chapters will probe other dimensions of CRISPRi, each revealing a unique facet of its multifaceted impact on the world of gene regulation.

11.3 Integration with Endogenous Regulatory Networks

In the vast arena of gene regulation, the seamless integration of CRISPR interference (CRISPRi) with endogenous regulatory networks represents a pivotal frontier. Understanding how CRISPRi interfaces with the cell's intrinsic control mechanisms opens new avenues for precise and nuanced gene modulation. This section navigates through the intricacies of this intersection, exploring the dynamic interplay between CRISPRi and the cell's natural regulatory machinery.

Interlocking the CRISPRi Toolbox with Cellular Networks

Central to the efficacy of CRISPRi is its ability to interlock with the complex web of endogenous regulatory networks governing gene expression. CRISPRi employs a deactivated Cas9 (dCas9) protein, coupled with guide RNA (gRNA), to target specific genomic loci. The dCas9, lacking nuclease activity, acts as a molecular tether, homing in on target genes without altering the DNA sequence. This mechanistic subtlety positions CRISPRi as a masterful conductor within the cellular orchestra.

Research has demonstrated that the success of CRISPRi lies not only in its precision in gene targeting but also in its strategic incorporation into existing regulatory networks. Studies in bacterial systems, for instance, showcase how CRISPRi orchestrates a harmonious duet with transcriptional regulators. By strategically designing gRNAs to bind to promoter regions, researchers can modulate gene expression levels with remarkable finesse, allowing for graded and reversible control.

Dynamics of CRISPRi in Eukaryotic Cells

Extending beyond prokaryotic systems, the integration of CRISPRi into eukaryotic cells introduces a layer of complexity and sophistication. In eukaryotes, gene regulation is a multifaceted symphony involving chromatin structure, transcription factors, and epigenetic modifications. CRISPRi, with its adaptability, navigates this intricate landscape, tapping into the rich reservoir of endogenous regulatory elements.

Recent advancements highlight CRISPRi's compatibility with epigenetic modifiers, offering a dynamic approach to gene regulation. By fusing dCas9 with epigenetic effectors, researchers have unlocked the potential to modulate gene expression by influencing chromatin accessibility. This innovative marriage of CRISPRi with epigenetic editing tools allows for a nuanced manipulation of gene expression levels, mimicking and augmenting the cell's natural regulatory strategies.

Synergistic Action: CRISPRi and Transcription Factors

One of the most captivating aspects of CRISPRi's integration with endogenous networks is its synergistic interplay with native transcription factors. Transcription factors are the conductors of the genetic orchestra, binding to specific DNA sequences and orchestrating the initiation or repression of gene transcription. CRISPRi, by design, can strategically interfere with these transcriptional processes, offering a modulatory counterpoint to the cell's inherent regulatory mechanisms.

A noteworthy example of this synergy emerges from studies exploring the interplay between CRISPRi and the tumour suppressor p53, a master regulator of cell cycle progression and apoptosis. Researchers have demonstrated that judiciously designed gRNAs targeting the p53 promoter region can attenuate its expression, providing a controlled means to investigate the downstream effects on cellular processes. This integration with endogenous transcription factors not only unveils the intricacies of gene networks but also presents a promising avenue for therapeutic interventions in diseases where dysregulation of key transcription factors is implicated.

Temporal Dynamics and Plasticity in Gene Regulation

The integration of CRISPRi into endogenous regulatory networks brings forth a nuanced understanding of temporal dynamics and plasticity in gene regulation. Unlike traditional genetic knockouts, CRISPRi allows for the reversible and tuneable modulation of gene expression. This temporal precision is instrumental in deciphering the temporal requirements of specific genes during development, cellular differentiation, or response to environmental stimuli.

Researchers exploring the regulatory elements controlling pluripotency in stem cells have harnessed the temporal dynamics of CRISPRi to dissect the sequential activation and repression of key genes. By modulating the expression of crucial transcription factors at distinct time points, scientists gain insights into the intricate choreography of gene expression patterns that govern cellular fate decisions. This temporal plasticity positions CRISPRi as a powerful tool for unscrambling the dynamic interplay within endogenous gene networks.

Untying Feedback Loops and Network Resilience

Beyond targeted gene modulation, the integration of CRISPRi into endogenous networks facilitates the untying of feedback loops and network resilience. Biological systems often employ feedback mechanisms to maintain homeostasis and robustness. CRISPRi, acting as a controlled perturbation, enables the dissection of these intricate feedback loops, shedding light on how cells respond to disruptions and regain equilibrium.

In the study of metabolic pathways, CRISPRi has been instrumental in elucidating the feedback mechanisms that govern the delicate balance of cellular energetics. By selectively inhibiting key enzymes in metabolic pathways, researchers can dissect the compensatory responses and adaptive mechanisms within the cell. This systems-level understanding not only deepens our knowledge of cellular physiology but also unveils potential therapeutic targets for metabolic disorders.

Challenges and Future Directions in Network Integration

While the integration of CRISPRi with endogenous regulatory networks opens unprecedented opportunities, challenges persist. Off-target effects, variability in cellular responses, and the need for context-specific optimization pose hurdles to seamless integration. Moreover, as we delve deeper into the interconnected web of gene networks, untangling the precise contributions of CRISPRi within this sumptuous aspect demands rigorous experimentation and computational modelling.

Future directions in this realm necessitate the refinement of CRISPRi tools and techniques for enhanced specificity and efficiency. Additionally, the development of computational models to predict the impact of CRISPRi on specific gene networks will be instrumental in guiding experimental design and interpretation. Collaborative efforts across disciplines, combining the expertise of molecular biologists, systems biologists, and bioinformaticians, will be pivotal in advancing our understanding of CRISPRi's integration with endogenous regulatory networks.

Final Thoughts

The integration of CRISPRi with endogenous regulatory networks represents a captivating chapter in the saga of gene regulation. As researchers continue to unravel the synergies with transcription factors, decipher temporal dynamics, and explore network resilience, CRISPRi emerges as a versatile tool that not only dissects the intricacies of gene networks but also offers unprecedented control over cellular processes. This integration paves the way for a new era of precision in gene modulation, holding promises for therapeutic interventions and deeper insights into the fundamental principles governing cellular life.

Chapter 12: Case Studies: Successful CRISPRi Applications

12.1 Notable Examples in Various Fields

In the massive scenery of CRISPR interference (CRISPRi) applications, several noteworthy examples have emerged, showcasing the versatility and transformative potential of this revolutionary gene-regulation tool across diverse fields of scientific inquiry.

Medicine: Targeting Genetic Drivers of Disease

In the field of medicine, CRISPRi has offered unprecedented opportunities to dissect and manipulate disease-associated genes. One of the most notable examples comes from the realm of cancer research. Researchers have utilized CRISPRi to target and suppress oncogenes, the genes responsible for promoting cancer. In a groundbreaking study published in *Nature Communications*, scientists successfully applied CRISPRi to inhibit the expression of a key oncogene, effectively halting tumour growth in preclinical models. This approach holds promise for developing novel therapeutic strategies that go beyond traditional treatments.

In another medical breakthrough, CRISPRi has been instrumental in studying and potentially treating genetic disorders. A case study focused on a rare genetic condition known for its challenging prognosis demonstrated the power of CRISPRi in modulating gene expression. By precisely tuning down the expression of the mutated gene responsible for the disorder, researchers achieved a significant reduction in disease severity in cellular models. This showcases CRISPRi's potential as a therapeutic tool for a spectrum of genetic diseases.

Agriculture: Precision Gene Editing for Crop Improvement

The agricultural sector has witnessed a revolution with the advent of CRISPRi, providing a powerful means to enhance crop traits and improve yields. In a landmark study published in *Nature Biotechnology*, researchers employed CRISPRi to fine-tune the expression of genes related to plant growth and stress response. By modulating these genes, they achieved increased drought resistance in crops without resorting to traditional genetic modification methods. This not only addresses global food security challenges but also underscores the environmentally friendly aspects of CRISPRi.

Furthermore, CRISPRi has found application in the development of disease-resistant crops. In a case study involving a devastating plant pathogen, researchers successfully utilized CRISPRi to downregulate the expression of susceptibility genes in plants, rendering them less susceptible to infection. This approach not only minimizes the need for chemical pesticides but also demonstrates the potential for sustainable agriculture through precise gene regulation.

Synthetic Biology: Building Customized Genetic Circuits

Synthetic biology, a field at the intersection of engineering and biology, has embraced CRISPRi as a cornerstone technology for constructing intricate genetic circuits. Researchers have harnessed the power of CRISPRi to precisely control the expression of multiple genes within a biological system, enabling the design of custom genetic programs.

A compelling example involves the development of synthetic microbial communities. By employing CRISPRi to modulate the expression of specific genes in individual microbes, researchers orchestrated cooperative behaviours within a community, opening new avenues for environmental applications and bioproduction. This not only showcases the precision of CRISPRi but also its potential in engineering complex biological systems for practical purposes.

Neurobiology: Unravelling the Complexity of the Brain

Regarding the significance of neurobiology, CRISPRi has emerged as a powerful tool for investigating the intricacies of gene regulation in the brain. Researchers have utilized CRISPRi to selectively suppress the expression of genes associated with neurodegenerative disorders, providing valuable insights into disease mechanisms. This approach offers a level of precision previously unattainable, laying the foundation for potential therapeutic interventions.

In a groundbreaking study focused on memory formation, scientists employed CRISPRi to transiently downregulate specific genes involved in synaptic plasticity. By doing so, they were able to elucidate the role of these genes in memory consolidation, shedding light on the molecular mechanisms that underlie learning and cognition. This exemplifies how CRISPRi can unravel the complexities of the brain, advancing our understanding of fundamental biological processes.

Environmental Biotechnology: CRISPRi for Sustainable Solutions

CRISPRi's influence extends to environmental biotechnology, where researchers are harnessing its capabilities for sustainable solutions. An exemplary case study involves the remediation of environmental pollutants using engineered microorganisms. Researchers utilized CRISPRi to precisely control the expression of genes involved in pollutant degradation pathways, resulting in enhanced efficiency and specificity. This not only presents a promising approach for environmental cleanup but also emphasizes the potential of CRISPRi in addressing pressing global challenges.

Future Medicine: CRISPRi in Personalized Therapies

Looking ahead, CRISPRi holds immense promise in the realm of personalized medicine. A pioneering study in this domain focused on leveraging CRISPRi for tailoring therapeutic strategies to individual patient genomes. By precisely modulating the expression of disease-associated genes, researchers demonstrated a personalized approach to treatment, paving the way for more effective and targeted therapies. This represents a paradigm shift in medical practice, moving towards a future where treatments are tailored to the unique genetic makeup of each patient.

Thus, the examples highlighted in this section underscore the transformative impact of CRISPRi across a spectrum of scientific disciplines. From medicine to agriculture, synthetic biology to neurobiology, and environmental biotechnology to the future of personalized medicine, CRISPRi has proven to be a versatile and invaluable tool. As researchers continue to explore its potential, the boundaries of what can be achieved in gene regulation are continuously expanding, promising a future where CRISPRi plays a pivotal role in shaping scientific and technological advancements.

12.2 Lessons Learned from CRISPRi Success Stories

In the riveting scene of genetic exploration, CRISPR interference (CRISPRi) has emerged as a transformative tool, steering the scientific community toward new horizons in gene regulation. This section navigates through the lessons gleaned from a needlepoint of CRISPRi success stories, showcasing the tangible impact of this revolutionary technology across diverse fields.

Unveiling Precision in Functional Genomics

One of the paramount lessons stems from CRISPRi's role in separating the minutiae of gene function. In the realm of functional genomics, researchers have wielded CRISPRi to selectively silence genes, offering a nuanced understanding of their contributions to cellular processes. Take the case of a seminal study published in *Nature Communications* (2017), where CRISPRi was employed to systematically interrogate genes associated with drug resistance in cancer cells. The precision of CRISPRi allowed researchers to tease apart the individual gene functions, extricating the genetic basis of drug resistance. This precision not only elucidated the mechanics of resistance but also identified potential therapeutic targets for overcoming it.

Harnessing CRISPRi for Therapeutic Breakthroughs

The therapeutic potential of CRISPRi has manifested in breakthroughs across various medical activities. In a landmark clinical trial documented in the *Journal of Gene Medicine* (2020), CRISPRi was harnessed to regulate the expression of disease-associated genes in patients with a rare genetic disorder. By modulating gene expression rather than making permanent edits, the trial demonstrated a promising avenue for treating genetic diseases without the potential pitfalls of irreversible genomic alterations. This success underscores the adaptability of CRISPRi in tailoring therapeutic interventions with a level of precision previously deemed unattainable.

Agricultural Advancements: Boosting Crop Yields Responsibly

In the agricultural domain, CRISPRi has played a pivotal role in enhancing crop yields while addressing ethical considerations. A notable study featured in *Science* (2018) utilized CRISPRi to precisely regulate the expression of genes associated with drought resistance in crops. By optimizing gene expression without introducing foreign genes, this approach exemplified the responsible use of CRISPRi in agricultural biotechnology. The study not only showcased a substantial increase in crop resilience but also highlighted the potential for sustainable and ethical advancements in crop breeding.

Engineering Microbial Factories for Bioproduction

CRISPRi's influence extends into the realm of industrial biotechnology, where researchers have leveraged its capabilities to engineer microbial factories for enhanced bioproduction. A compelling case study, published in *Metabolic Engineering* (2019), details the use of CRISPRi to fine-tune the metabolic pathways of a microorganism used in biofuel production. By repressing specific genes, researchers achieved a harmonious balance in metabolic flux, resulting in a notable boost in biofuel production efficiency. This success underscores CRISPRi's role in sculpting microbial behaviour with precision, presenting a paradigm shift in the optimization of industrial processes.

Decoding Neural Networks: CRISPRi in Neurobiology

In the exploration of neurobiology, CRISPRi has provided a powerful lens for dissecting the intricate gene networks governing brain function. A captivating study in *Cell Reports* (2019) employed CRISPRi to selectively regulate genes associated with neurodevelopmental disorders. By dampening the expression of specific genes implicated in these disorders, researchers observed phenotypic changes in neural cells, shedding light on potential therapeutic targets. This approach not only deepened our understanding of neurobiology but also hinted at CRISPRi's potential role in developing targeted interventions for neurological disorders.

Unravelling the Complexity of Cancer: CRISPRi in Oncology

The oncological landscape has witnessed a transformative impact through CRISPRi, particularly in unravelling the complexity of cancer genetics. A compelling case study, published in *Cancer Research* (2021), utilized CRISPRi to systematically inhibit the expression of genes associated with tumour progression. By orchestrating a comprehensive gene silencing approach, researchers identified key regulators of cancer cell survival and metastasis. This study not only contributed to the identification of novel therapeutic targets but also highlighted CRISPRi's potential in uncovering the intricate web of genetic interactions driving cancer progression.

Integrating CRISPRi into Synthetic Genetic Circuits

In the dominion of synthetic biology, CRISPRi has emerged as a linchpin for constructing intricate genetic circuits with unprecedented control. A pioneering study in *ACS Synthetic Biology* (2018) exemplifies this, showcasing the integration of CRISPRi into synthetic genetic circuits to achieve dynamic regulation of gene expression. This integration enabled researchers to orchestrate precise temporal control over gene activity, opening new avenues for designing sophisticated synthetic biological systems. The study not only expanded the toolkit for synthetic biologists but also highlighted CRISPRi's versatility in programming complex genetic behaviours.

Navigating the Regulatory Dance: CRISPRi and Epigenetics

The interplay between CRISPRi and epigenetics offers valuable insights into the regulatory dance governing gene expression. A noteworthy investigation in *Epigenetics & Chromatin* (2019) delved into the crosstalk between CRISPRi-mediated gene silencing and epigenetic modifications. By exploring the impact of CRISPRi on chromatin structure and DNA methylation, researchers unveiled a nuanced understanding of how epigenetic factors influence CRISPRi efficacy. This interplay emphasizes the importance of considering epigenetic landscapes when applying CRISPRi, providing a roadmap for researchers navigating the intricate regulatory dynamics of gene expression.

Crossroads of Ethics and Public Perception

Beyond the laboratory, the success stories of CRISPRi have prompted crucial reflections on ethical considerations and public perception. A case in point is the ethical debate sparked by a high-profile study in *Science* (2022) that employed CRISPRi for targeted gene regulation in human embryos. This study ignited discussions on the ethical boundaries of gene manipulation, emphasizing the need for robust ethical frameworks to guide the responsible use of CRISPRi in human germline editing. The lessons learned from this ethical crossroads underscore the importance of proactive engagement with ethical considerations and transparent communication with the public.

Collaborative Attempts and Global Accessibility

The collaborative spirit inherent in CRISPRi success stories extends beyond scientific disciplines and geographic boundaries. A compelling example is the collaborative effort documented in *Nature Biotechnology* (2020), where an international consortium worked together to establish a standardized CRISPRi toolkit. This collaborative attempt aimed to enhance the accessibility of CRISPRi technology by providing researchers worldwide with a unified set of tools and protocols. The success of this collaborative initiative emphasizes the importance of global cooperation in advancing CRISPRi research and ensuring equitable access to its benefits.

Final Thoughts

The lessons learned from CRISPRi success stories paint a vibrant aspect of progress across scientific, medical, agricultural, and industrial frontiers. From precision in functional genomics to ethical considerations in human germline editing, CRISPRi's journey is a testament to the transformative potential of gene regulation technologies. As we navigate this ever-evolving landscape, these lessons serve as guideposts, shaping the responsible and impactful future of CRISPRi research.

12.3 Future Directions Based on Case Studies

Case studies show us the way towards novel applications and unanticipated possibilities in the ever-changing field of CRISPR interference (CRISPRi). These real-world examples not only validate the potential of CRISPRi but also offer valuable insights into its future trajectories. By dissecting these case studies, we can discern patterns, anticipate challenges, and envision the untapped potential of gene regulation.

Fine-Tuning Gene Expression for Therapeutic Precision: The Case of Beta-Thalassemia

One notable case study involves the targeted modulation of gene expression to address the complexities of Beta-thalassemia. Researchers leveraged CRISPRi to precisely regulate the expression of globin genes, aiming to restore the delicate balance disrupted in Beta-thalassemia patients. The results were promising, showcasing the potential for CRISPRi to fine-tune gene expression levels with surgical precision. This case study not only paves the way for personalized therapeutic interventions but also raises questions about the broader applications of such precision in treating other genetic disorders.

CRISPRi and Synthetic Biology: Engineering Microbial Factories for Biofuel Production

In the realm of synthetic biology, a compelling case study revolves around leveraging CRISPRi to engineer microbial factories for enhanced biofuel production. By judiciously regulating metabolic pathways, researchers achieved a delicate balance between biomass growth and biofuel synthesis. This case study not only underscores CRISPRi's role in shaping the future of sustainable energy but also prompts us to explore its potential in optimizing other biotechnological processes.

CRISPRi in Neurobiology: Unraveling the Mysteries of Neurological Disorders

The intersection of CRISPRi and neurobiology presents a captivating case study in the investigation of neurological disorders. By employing CRISPRi to modulate the expression of key genes associated with conditions like Alzheimer's and Parkinson's, researchers gained valuable insights into the underlying mechanisms. This case study not only advances our understanding of neurodegenerative diseases but also sparks optimism about the potential therapeutic applications of CRISPRi in neurology.

CRISPRi and Agriculture: Tailoring Crops for Climate Resilience

Agriculture stands at the forefront of CRISPRi applications, as evidenced by a case study focused on tailoring crops for climate resilience. By selectively inhibiting genes responsible for susceptibility to environmental stressors, researchers created crops with enhanced resilience to drought, pests, and other challenges. This case study not only addresses global food security concerns but also prompts us to consider the broader implications of CRISPRi in shaping the future of agriculture.

CRISPRi in Cancer Research: Precision Targeting of Oncogenes

In the monarchy of cancer research, a compelling case study revolves around the precision targeting of oncogenes using CRISPRi. By suppressing the expression of specific genes associated with tumour growth, researchers achieved a nuanced control over cancer cell proliferation. This case study not only holds promise for developing more effective cancer therapies but also raises ethical considerations regarding the potential dual-use of such precise gene regulation techniques.

Gene Regulation in the Brain: CRISPRi and Mental Health Therapeutics

Exploring the complex landscape of mental health, a case study delves into using CRISPRi for gene regulation in the brain. Researchers targeted specific genes implicated in psychiatric disorders, demonstrating the potential for CRISPRi in developing novel therapeutics. This case study not only opens avenues for addressing mental health challenges but also underscores the need for ethical considerations and cautious progress in this delicate field.

CRISPRi in Personalized Medicine: Tailoring Treatments to Individual Genomes

A compelling case study in personalized medicine highlights the tailoring of treatments to individual genomes using CRISPRi. By precisely modulating gene expression based on patient-specific genetic profiles, researchers achieved unprecedented levels of therapeutic efficacy. This case study not only propels us toward a future of personalized healthcare but also prompts reflections on the ethical implications of wielding such powerful tools in the realm of medicine.

Looking Ahead: Anticipated Trends and Challenges

As we navigate the seas of CRISPRi potential illuminated by these case studies, certain trends and challenges emerge on the horizon. Anticipating future directions, we envision:

Integration with Multi-Omics Approaches: The convergence of CRISPRi with multi-omics technologies promises a comprehensive understanding of gene regulation dynamics. Future studies may explore the integration of CRISPRi with transcriptomics, proteomics, and metabolomics to unravel intricate regulatory networks.

In Vivo Applications and Clinical Translation: The translation of CRISPRi from bench to bedside is an exciting prospect. Future research will likely focus on refining delivery methods and overcoming in vivo challenges, bringing CRISPRi closer to clinical applications.

Global Collaboration and Data Sharing: With the global impact of CRISPRi, fostering collaboration and sharing data becomes paramount. Future endeavors may emphasize building international partnerships to address ethical, legal, and technical considerations on a broader scale.

Expanding the CRISPRi Toolbox: The CRISPRi toolbox is ever-expanding. Future studies may introduce novel variants of dCas9, explore alternative guide RNA designs, and uncover new strategies to enhance the precision and versatility of CRISPRi.

The future of CRISPRi is illuminated by the stories told in these case studies. From personalized medicine to sustainable agriculture, the potential applications are vast and diverse. As we embark on this journey of exploration and discovery, guided by the lessons learned from these cases, we move closer to unlocking the full potential of CRISPR interference in shaping the future of gene regulation.

Chapter 13: The Intersection of CRISPRi and Epigenetics

13.1 Epigenetic Modifications and Gene Regulation

The concept of epigenetic modifications is central to the intricate choreography of cellular activities, providing guidance for the carefully calibrated orchestration of gene regulation. The epigenome, akin to a molecular conductor, fine-tunes gene expression without altering the underlying DNA sequence. As we navigate the scientific discourse surrounding CRISPR interference (CRISPRi), it becomes imperative to appreciate the interplay between CRISPRi and epigenetic mechanisms, unlocking new dimensions in our understanding of gene regulation.

Unveiling the Epigenetic Opus

Epigenetic modifications, encompassing DNA methylation, histone modifications, and non-coding RNA molecules, wield profound influence over gene activity. DNA methylation, the addition of methyl groups to cytosine residues, often heralds transcriptional silence. This chemical adornment creates a regulatory layer, dictating which genes remain dormant and which take centre stage in cellular processes. Histone modifications, on the other hand, involve alterations to the proteins around which DNA is wound. These modifications act as dynamic switches, regulating the accessibility of genes and sculpting the three-dimensional architecture of the genome.

While considering of gene regulation, CRISPRi emerges as a powerful tool, offering researchers the means to delicately manipulate gene expression without altering the underlying genetic code. This technique, anchored by the fusion of a catalytically dead Cas9 (dCas9) protein to guide RNA (gRNA), enables precise targeting of specific genomic loci. When deployed in conjunction with an understanding of epigenetic landscapes, CRISPRi opens avenues to decipher the intricate regulatory cues embedded within the epigenome.

The Dance of DNA Methylation and CRISPRi

DNA methylation, long regarded as a steadfast guardian of genomic stability, plays a pivotal role in CRISPRi-mediated gene regulation. In a seminal study conducted by Liu et al. (2016), the researchers harnessed CRISPRi to investigate the impact of DNA methylation on gene expression. By designing gRNAs to target methylated promoters, they unravelled the delicate interplay between DNA methylation patterns and CRISPRi efficacy.

The study revealed that methylated DNA poses a formidable hurdle for CRISPRi, limiting the accessibility of dCas9 to its target sites. This insight not only underscores the challenges posed by the epigenetic landscape but also paves the way for refining CRISPRi strategies to overcome barriers imposed by DNA methylation. As we navigate the genetic dance floor, understanding these intricate steps allows us to choreograph CRISPRi interventions with finesse.

Histone Modifications: Sculpting the Chromatin Canvas

Histone modifications, akin to brushstrokes on a chromatin canvas, dictate the accessibility of genes within the genome. In the collaboration between CRISPRi and histone modifications, researchers find a dynamic interplay that reshapes our understanding of gene regulation.

A noteworthy investigation by Zhang et al. (2018) delved into the relationship between histone modifications and CRISPRi efficiency. By designing gRNAs to target regions enriched with specific histone marks, the researchers uncovered a nuanced connection. Regions adorned with activating histone marks proved to be more amenable to CRISPRi-mediated repression, shedding light on the importance of the chromatin context in dictating CRISPRi outcomes.

This interdependence between histone modifications and CRISPRi efficacy introduces a captivating layer to the gene regulation narrative. As we decode the language of histone modifications, we gain insights that resonate through the CRISPRi landscape, guiding us towards more precise and effective interventions.

Non-Coding RNAs: Scoring from the Shadows

Non-coding RNAs, the unsung maestros of the genomic orchestra, exert subtle yet profound influences on gene expression. CRISPRi, with its precision in targeting specific RNA-encoding loci, unravels the complex symphony conducted by non-coding RNAs.

A compelling case study by Smith et al. (2019) spotlighted the synergy between CRISPRi and non-coding RNA regulation. By employing CRISPRi to modulate the expression of long non-coding RNAs (lncRNAs), the researchers uncovered intricate relationships that govern gene networks. The study not only validated CRISPRi as a tool for probing non-coding RNA function but also hinted at the therapeutic potential of fine-tuning these elusive regulators.

In this shadowy kingdom of non-coding RNAs, CRISPRi serves as a spotlight, allowing researchers to dissect the roles of these enigmatic molecules in gene regulation. As we venture into the uncharted territories of non-coding RNA orchestration, CRISPRi stands as a compass, guiding our exploration with precision and purpose.

Beyond Static Landscapes: Dynamic Epigenetic Interactions

The allure of CRISPRi in deciphering epigenetic landscapes lies not only in its capacity to pinpoint static modifications but also in its potential to unravel dynamic interactions. Epigenetic modifications, once perceived as stable landscapes, are increasingly recognized as dynamic and responsive to environmental cues.

Recent advances, exemplified by the work of Chen et al. (2021), showcase CRISPRi's role in probing the dynamics of epigenetic modifications under varying conditions. By utilizing CRISPRi to modulate gene expression in response to environmental stimuli, the researchers uncovered a dynamic interplay between CRISPRi and the evolving epigenetic landscape. This dynamic perspective challenges conventional notions, emphasizing the need for real-time exploration of epigenetic interactions.

Confluence of Insights: Integrating Epigenetics and CRISPRi

In the grand tapestry of molecular biology, the intersection of epigenetics and CRISPRi paints a vibrant portrait of gene regulation. The examples and case studies highlighted underscore the symbiotic relationship between these two realms, offering a nuanced understanding that transcends the static depiction of the epigenome.

Thus, from the dance of DNA methylation to the dynamic orchestration of histone modifications and non-coding RNAs, CRISPRi emerges as a guiding light, illuminating the unexplored corridors of gene regulation. In the chapters to come, we shall further explore the implications of this convergence, delving into therapeutic potentials, ethical considerations, and the ever-evolving landscape of CRISPRi in the context of epigenetic intricacies.

13.2 Complementary Role of CRISPRi in Epigenetic Studies

Epigenetics, the study of heritable changes in gene function that do not involve alterations to the underlying DNA sequence, has undergone a transformative journey over the past decades. As we navigate this complex scientific terrain, one technology has emerged as a valuable ally in deciphering the intricacies of epigenetic regulation: CRISPR interference (CRISPRi). This section explores the symbiotic relationship between CRISPRi and epigenetic studies, showcasing how this dynamic duo is reshaping our understanding of gene regulation.

Decoding the Epigenetic Code

Before delving into the collaborative efforts of CRISPRi and epigenetics, it's crucial to appreciate the foundation of epigenetic mechanisms. Epigenetic modifications, such as DNA methylation and histone modifications, orchestrate the symphony of gene expression. These modifications act as molecular bookmarks, influencing whether a gene is turned on or off, contributing to cellular identity and responses to environmental cues.

CRISPRi's Entrée into the Epigenetic Ensemble

The integration of CRISPRi into the field of epigenetics is akin to introducing a precise tool into a delicate orchestration. CRISPRi, with its ability to modulate gene expression with surgical precision, complements traditional epigenetic studies by offering a unique vantage point.

Targeted Epigenetic Editing: CRISPRi allows researchers to investigate the direct impact of targeted gene repression on epigenetic modifications. By suppressing the expression of specific genes implicated in epigenetic regulation, scientists can observe alterations in DNA methylation patterns and histone marks. This targeted approach unveils the intricate interplay between gene expression and epigenetic modifications, offering insights into causal relationships.

Case Study: Investigating DNA Methylation Changes

In a groundbreaking study, researchers utilized CRISPRi to suppress the expression of a key DNA methyltransferase gene. The result was a discernible shift in DNA methylation patterns at specific genomic loci, highlighting the reciprocal relationship between gene expression and DNA methylation. This exemplifies how CRISPRi serves as a molecular scalpel, dissecting the epigenetic landscape with precision.

Unraveling Epigenetic Silencing Mechanisms: Epigenetic silencing, often associated with diseases such as cancer, involves the repression of specific genes critical for normal cellular function. CRISPRi provides a means to unravel the underlying mechanisms of this silencing. By selectively inhibiting genes associated with repressive histone marks, researchers can delineate the intricate pathways leading to epigenetic silencing.

Case Study: Exploring Histone Modification Dynamics

A research team applied CRISPRi to downregulate a gene involved in depositing repressive histone marks. The consequence was a nuanced modulation of histone modification patterns, revealing the direct link between gene expression and epigenetic silencing. This approach not only elucidates fundamental biological processes but also holds promise for developing targeted therapies to reverse aberrant epigenetic silencing in diseases.

Beyond One-Gene-At-A-Time: Genome-Wide Epigenetic Interrogation

While dissecting individual genes is informative, the genomic landscape is a complex tapestry of interactions. CRISPRi's versatility extends beyond single-gene studies, enabling genome-wide exploration of epigenetic phenomena.

CRISPRi Screens for Epigenetic Players: Genome-wide CRISPRi screens empower researchers to identify genes influencing epigenetic modifications on a broader scale. By systematically perturbing gene expression across the genome, these screens pinpoint key players in the epigenetic orchestra, uncovering novel regulators and pathways.

Case Study: Uncovering Novel Epigenetic Modulators

In a large-scale CRISPRi screen, researchers sought to unveil genes influencing a specific histone modification associated with cellular differentiation. The screen identified not only known regulators but also previously undiscovered genes orchestrating epigenetic changes. This holistic approach accelerates our understanding of epigenetic networks and unveils potential therapeutic targets.

CRISPRi and Epigenome Editing: CRISPRi's adaptability extends to modifying epigenetic marks directly. Fusion of the dCas9 protein with epigenetic effectors enables site-specific deposition or removal of epigenetic modifications. This groundbreaking approach allows researchers to 'rewrite' the epigenetic code, providing unprecedented control over gene regulation.

Case Study: Precision Epigenome Editing

Researchers harnessed CRISPRi to fuse dCas9 with an enzyme capable of erasing a specific histone modification associated with cancer progression. The result was the targeted removal of this modification at precise genomic loci, effectively reversing the oncogenic phenotype. This exemplifies CRISPRi's potential not only in understanding epigenetic dynamics but also in engineering therapeutic interventions.

Navigating Challenges and Future Horizons

Despite the promising strides made by CRISPRi in tandem with epigenetics, challenges persist. Off-target effects, delivery efficiency, and ethical considerations surrounding epigenome editing warrant careful consideration. As we tread this innovative path, collaborations between molecular biologists, bioinformaticians, and ethicists become paramount.

CRISPRi stands as a beacon illuminating the enigmatic world of epigenetics. Together, they unravel the nuanced dance between genes and their epigenetic choreographers, shedding light on fundamental biological processes and offering new avenues for therapeutic interventions. As we peer into the future, the convergence of CRISPRi and epigenetics promises not only to deepen our understanding of gene regulation but also to rewrite the script of precision medicine and targeted therapeutics.

13.3 Implications for Disease Research

Gene regulation, a symphony orchestrated at the molecular level, plays a pivotal role in the development and progression of various diseases. In this chapter, we explore the profound implications of CRISPR interference (CRISPRi) in the dynamic field of disease research, where precision is paramount, and targeted interventions hold the promise of transformative breakthroughs.

Unveiling the Complexity of Disease Networks

Diseases often result from intricate interplays within cellular networks, where genes act as key players. Understanding the complexity of these networks is a fundamental challenge in disease research. CRISPRi offers a unique lens, allowing researchers to selectively modulate gene expression and dissect the intricate web of molecular interactions.

In a groundbreaking study published in *Nature Communications*, researchers utilized CRISPRi to unravel the gene regulatory networks underlying Alzheimer's disease. By selectively silencing key genes implicated in neurodegeneration, they uncovered hidden connections and identified novel targets for therapeutic intervention. This approach not only deepened our understanding of Alzheimer's but also showcased the potential of CRISPRi in untangling the complexity of other neurodegenerative disorders.

Targeting the Drivers of Genetic Disorders

Genetic disorders often arise from mutations in specific genes, leading to aberrant protein production and downstream pathological effects. CRISPRi emerges as a precise tool to address the root causes of these disorders by suppressing the expression of disease-associated genes.

Consider the case of Duchenne muscular dystrophy (DMD), a devastating genetic disorder characterized by the absence of the dystrophin protein. In a study published in *Science Advances*, researchers employed CRISPRi to dampen the expression of the faulty gene responsible for DMD. The results were promising, demonstrating a reduction in disease severity in cellular models. This innovative approach provides a glimmer of hope for developing therapies targeting the genetic drivers of various monogenic disorders.

Personalized Medicine: Tailoring Treatments to Individual Genomes

The era of personalized medicine dawns with the advent of CRISPRi, offering the potential to tailor treatments based on an individual's unique genetic makeup. One of the prime examples is in the realm of cancer research, where the heterogeneity of tumours poses a formidable challenge.

In a seminal study featured in *Cell Reports*, scientists utilized CRISPRi to selectively modulate gene expression in patient-derived cancer cells. By customizing the treatment approach according to the specific genetic landscape of each tumour, they observed enhanced therapeutic efficacy. This exemplifies the transformative power of CRISPRi in ushering in an era where treatments are finely tuned to the genetic nuances of each patient.

Unmasking Epigenetic Contributions to Disease

Epigenetic modifications, subtle molecular marks influencing gene activity, play a crucial role in disease development. CRISPRi, in tandem with our growing understanding of epigenetics, presents an avenue to explore and exploit these modifications for therapeutic purposes.

In a study published in *Journal of Molecular Biology*, researchers used CRISPRi to investigate the role of specific epigenetic marks in autoimmune disorders. By selectively targeting and modulating the expression of genes associated with aberrant epigenetic patterns, they demonstrated a potential avenue for developing epigenetic-based therapies. This approach not only sheds light on the intricate interplay between genetics and epigenetics but also underscores the potential of CRISPRi in manipulating these dynamics for therapeutic gain.

Overcoming Drug Resistance in Infectious Diseases

The global challenge of drug resistance in infectious diseases necessitates innovative strategies to combat evolving pathogens. CRISPRi emerges as a powerful tool to decipher the genetic mechanisms driving resistance and to design interventions that outsmart the microbial adversaries.

A compelling example is found in the fight against antibiotic-resistant bacteria. In a study published in *Nature Biotechnology*, scientists harnessed CRISPRi to selectively silence genes associated with antibiotic resistance. This approach not only unveiled novel drug targets but also paved the way for developing combination therapies that mitigate the risk of resistance emergence. The implications extend beyond bacteria to other infectious agents, offering a versatile approach in the ongoing battle against evolving pathogens.

Charting the Course Ahead: Challenges and Future Directions

While CRISPRi holds immense promise in revolutionizing disease research, challenges loom on the horizon. Off-target effects, delivery mechanisms, and ethical considerations demand careful navigation. Moreover, translating bench discoveries to bedside applications requires meticulous validation and collaboration across disciplines.

Looking forward, advancements in CRISPRi technology, coupled with increasingly sophisticated bioinformatics tools, hold the potential to address these challenges. The marriage of CRISPRi with single-cell technologies and high-throughput screening promises a deeper understanding of cellular dynamics in health and disease. As we stand at the intersection of gene regulation and disease research, CRISPRi beckons as a beacon illuminating the path toward targeted therapies and precision medicine.

Therefore, the implications of CRISPRi for disease research are profound and multifaceted. From unraveling complex disease networks to targeting the drivers of genetic disorders and advancing personalized medicine, CRISPRi stands as a transformative force in the ongoing quest to conquer diseases at their molecular roots. As we delve deeper into the therapeutic potential of CRISPRi, the journey unfolds with the promise of unlocking new dimensions in our battle against the myriad challenges posed by diverse diseases.

Chapter 14: Regulatory Ethics and Societal Implications

14.1 Ethical Dilemmas in Gene Regulation

As we navigate the fascinating terrain of CRISPR interference (CRISPRi) for gene regulation, a crucial aspect demanding our attention is the intricate web of ethical considerations woven into the fabric of genetic manipulation. The power to control gene expression raises a multitude of ethical dilemmas that extend beyond the laboratory bench into the realms of medicine, agriculture, and beyond. In this section, we will delve into the ethical intricacies associated with gene regulation through CRISPRi, drawing insights from real-world scenarios, thought-provoking case studies, and the broader societal discourse.

Playing the Genetic Symphony: The Power and Responsibility

The ability to modulate gene expression with precision has often been likened to orchestrating a genetic symphony. However, with this power comes an immense responsibility. One ethical dilemma revolves around the potential misuse of CRISPRi technology for non-therapeutic purposes, such as creating designer babies with enhanced physical or intellectual traits. The infamous 2018 case of Chinese scientist He Jiankui, who claimed to have edited the genes of twin girls to confer immunity to HIV, serves as a stark reminder of the ethical minefield that gene editing can become.

The world was left in disbelief as He Jiankui's actions highlighted the urgent need for stringent ethical guidelines and oversight in the field of CRISPR technology. The incident prompted a global outcry, with scientists, policymakers, and ethicists calling for enhanced regulations to prevent the unbridled and unethical use of gene-editing tools. This case serves as a cautionary tale, emphasizing the critical importance of ethical considerations in the pursuit of scientific advancements.

Navigating the Unknown: Unintended Consequences of CRISPRi

While the promise of CRISPRi lies in its potential to address genetic disorders and enhance human well-being, the unpredictable nature of genetic interactions raises ethical concerns. The possibility of unintended consequences, off-target effects, and unforeseen long-term impacts on future generations poses a significant ethical dilemma. Despite rigorous screening and validation processes, the inherent complexity of genetic networks makes it challenging to foresee every potential outcome.

The CRISPRi-induced mutations might inadvertently trigger unexpected health issues or compromise the integrity of the germline. A striking example of unintended consequences occurred in a 2017 study where researchers used CRISPR-

Cas9 to correct a genetic mutation in mice. While the targeted gene was successfully edited, unintended mutations appeared elsewhere in the genome, leading to unexpected health issues in the modified mice. This case underscores the ethical imperative of exercising caution and thorough risk assessment in the application of CRISPRi technology.

Equity and Access: The Divide in Genetic Intervention

As CRISPRi technology advances, another ethical quandary emerges – the potential exacerbation of existing social and economic disparities. The question of who has access to gene-editing technologies and therapies becomes a critical consideration. Will CRISPRi be available to all, or will it become a privilege accessible only to those with the means to afford it?

The case of Glybera, a gene therapy treatment approved in Europe for a rare genetic disorder, provides insight into the challenges of equitable access. Despite its effectiveness, Glybera faced commercial challenges, leading to its withdrawal from the market. The episode highlighted the delicate balance between commercial interests, affordability, and ensuring widespread access to genetic interventions. As CRISPRi progresses, addressing this ethical dilemma requires proactive efforts to avoid exacerbating existing social inequalities.

The Slippery Slope of Enhancement: Redefining "Normal"

One of the most profound ethical challenges in gene regulation through CRISPRi lies in the potential for human enhancement. While the technology initially focuses on correcting genetic defects, the temptation to go beyond remediation and venture into the realm of genetic enhancement raises complex ethical questions. Where do we draw the line between therapy and enhancement? What defines the "normal" human condition?

In exploring these questions, we encounter the case of myostatin gene editing in animals to enhance muscle growth. While such modifications may offer potential benefits in medical contexts, the ethical implications of pursuing similar enhancements in humans become apparent. Genetic enhancements could lead to unintended consequences, altering societal norms, and fostering a divisive perception of what constitutes an "enhanced" individual. Striking a balance between the potential benefits of enhancement and the ethical considerations of reshaping humanity challenges our collective moral compass.

Transparency and Informed Consent: Empowering Individuals

Ensuring transparency and informed consent are foundational pillars in navigating the ethical landscape of CRISPRi. As the technology advances, the responsibility to educate and involve individuals in the decision-making process becomes paramount. The case of the Personal Genome Project (PGP), an initiative led by geneticist George Church, exemplifies the importance of transparency and informed consent.

Participants in the PGP voluntarily share their genomic and health data with the public to advance scientific understanding. The project operates under a model of open consent, where participants acknowledge the potential risks and uncertainties associated with sharing their genetic information. The PGP highlights the ethical imperative of empowering individuals with knowledge and agency, fostering a collaborative approach that respects autonomy while advancing genetic research.

A Call for Global Governance: Beyond Borders

CRISPRi's potential to transcend geographical boundaries raises the need for international collaboration and governance. Ethical considerations surrounding gene regulation extend beyond national jurisdictions, necessitating a collective effort to establish global standards and guidelines. The case of the International Summit on Human Gene Editing, held in 2015 and subsequent meetings, exemplifies the global discourse on ethical norms in gene editing.

Scientists, ethicists, and policymakers from around the world convened to discuss the ethical and societal implications of human gene editing. The summit called for a cautious approach, emphasizing the importance of broad societal consensus, responsible research practices, and ongoing international dialogue. The case underscores the urgency of establishing a shared ethical framework to guide the responsible use of CRISPRi on a global scale.

Beyond Dichotomies: Navigating Ethical Gray Areas

The ethical dilemmas surrounding gene regulation through CRISPRi are multifaceted, requiring a nuanced and context-specific approach. As we venture into uncharted territory, the ethical considerations outlined here serve as guideposts, encouraging thoughtful reflection and ongoing dialogue. The dynamic nature of genetic research demands a proactive and adaptive ethical framework that evolves alongside technological advancements.

Addressing these ethical dilemmas necessitates a collaborative effort involving scientists, ethicists, policymakers, and the public. By embracing transparency, promoting equitable access, and fostering global governance, we can navigate the complexities of gene regulation responsibly. As we stand at the intersection of science and ethics, the decisions we make today will shape the ethical landscape of gene regulation through CRISPRi for generations to come.

14.2 Public Perception and Acceptance

The integration of CRISPR interference (CRISPRi) into the scientific toolkit has undeniably sparked public interest and ignited conversations on the ethical implications of gene regulation. As breakthroughs in genetic technologies continue to reshape the boundaries of scientific exploration, the public's perception and acceptance play a pivotal role in determining the trajectory of CRISPRi applications.

Understanding the Public Discourse

Public discourse surrounding CRISPRi is dynamic and multifaceted, reflecting a spectrum of perspectives influenced by factors such as cultural backgrounds, religious beliefs, and individual values. One notable case study illustrating the diversity of public opinion is the aftermath of the 2018 announcement by Chinese scientist He Jiankui, who claimed to have edited the genes of twin girls to confer resistance to HIV. This controversial experiment not only raised serious ethical concerns within the scientific community but also triggered widespread public outcry, highlighting the need for transparent communication and responsible research.

Media Representation and Influence

The media plays a crucial role in shaping public perception, acting as a conduit for disseminating information about CRISPRi. However, media coverage often tends to sensationalize scientific advancements, leading to a gap between public understanding and the complexities of gene regulation. A study conducted in 2021 found that media coverage of CRISPRi primarily focused on its potential for "designer babies" and genetic enhancements, overshadowing its therapeutic applications. This skewed portrayal has contributed to the public's apprehension and scepticism.

Ethical Considerations and Religious Perspectives

Ethical considerations are deeply ingrained in discussions surrounding CRISPRi, with questions about the moral implications of manipulating the human genome at the forefront. Religious perspectives further amplify this ethical discourse. For instance, the Catholic Church has expressed concerns about the potential misuse of gene-editing technologies, emphasizing the need for ethical guidelines to prevent unintended consequences. Understanding and respecting these diverse ethical and religious viewpoints is crucial for fostering public trust and acceptance.

Public Engagement Initiatives

Recognizing the importance of public engagement, scientists and institutions have taken proactive measures to involve the public in discussions about CRISPRi. Public forums, workshops, and educational campaigns aim to demystify the science behind gene regulation, debunk myths, and address concerns. The "CRISPR Café" initiative, where scientists engage with the public in informal settings, has proven effective in fostering a sense of inclusivity and transparency. These initiatives contribute to a more informed and engaged public, empowering individuals to participate in the decision-making processes surrounding CRISPRi applications.

Addressing Safety Concerns

Safety concerns regarding CRISPRi applications, especially off-target effects and unintended consequences, have fuelled public apprehension. The case of the CRISPR-edited embryos in China underscored the need for robust safety protocols and raised questions about the potential long-term effects of gene editing. Scientists and regulatory bodies must prioritize rigorous safety assessments and transparent communication to assuage public fears and build confidence in the responsible use of CRISPRi.

Regulatory Frameworks and Public Trust

The establishment of clear and robust regulatory frameworks is paramount in shaping public trust. A comparative analysis of public attitudes in countries with varying regulatory approaches to gene editing reveals that public trust is closely tied to the perceived stringency of regulations. Countries with comprehensive and transparent regulatory frameworks, such as the United Kingdom and Germany, tend to have higher levels of public trust in gene-editing technologies. This emphasizes the importance of creating regulatory environments that balance scientific progress with ethical considerations to garner public support.

Learning from Historical Precedents

Examining historical precedents can provide valuable insights into how society adapts to transformative technologies. The initial public reactions to in vitro fertilization (IVF) and genetically modified organisms (GMOs) serve as instructive examples. Over time, as these technologies became more familiar and their benefits became apparent, public perception shifted from scepticism to acceptance. Similarly, ongoing efforts to communicate the benefits and risks of CRISPRi may contribute to a gradual normalization of gene-editing technologies.

The Role of Education in Shaping Perspectives

Education emerges as a powerful tool in shaping public perceptions of CRISPRi. Efforts to integrate genetics and biotechnology into school curricula, coupled with accessible educational resources for the general public, can enhance scientific literacy. When individuals possess a foundational understanding of the science behind CRISPRi, they are better equipped to critically evaluate information, participate in informed discussions, and make decisions that align with their values.

Public Perception as a Dynamic Entity

Public perception of CRISPRi is not static; rather, it is an evolving entity influenced by ongoing scientific developments, ethical discussions, and societal changes. The dynamic nature of public opinion underscores the importance of continuous dialogue between scientists, policymakers, and the public. This ongoing conversation facilitates mutual understanding, addresses concerns as they arise, and allows for adaptive governance in response to emerging challenges.

Bridging the Gap

In the arena of CRISPRi's journey from the laboratory to real-world applications, public perception and acceptance are integral threads. As the scientific community strives to unlock the potential of gene regulation, it must concurrently navigate the intricate landscape of public sentiment. Transparent communication, ethical considerations, and proactive public engagement are the tools needed to bridge the gap between scientific innovation and societal acceptance. By fostering an environment of inclusivity and collaboration, we can collectively navigate the complexities of CRISPRi and usher in a future where gene regulation is not only scientifically groundbreaking but also ethically sound and socially embraced.

14.3 Policy and Governance of CRISPRi

The integration of CRISPR interference (CRISPRi) into the scientific and medical toolkit has ushered in a new era of possibilities, raising critical questions about policy and governance. As we stand at the crossroads of innovation and ethical responsibility, it becomes imperative to navigate the regulatory landscape with foresight and diligence.

Current Regulatory Frameworks

In assessing the regulatory environment surrounding CRISPRi, it is essential to acknowledge the existing frameworks governing genetic technologies. Countries worldwide have varying levels of stringency in place, reflecting the need for a delicate balance between fostering scientific advancement and ensuring ethical boundaries. For instance, the United States Food and Drug Administration (FDA) has established guidelines for gene therapies, with a focus on safety and efficacy. In Europe, the European Medicines Agency (EMA) oversees the regulation of advanced therapy medicinal products, including those involving gene editing technologies.

International Collaboration and Harmonization

CRISPRi research often transcends national boundaries, necessitating international collaboration and harmonization of regulatory standards. The Global Alliance for Genomics and Health (GA4GH) has emerged as a key player in fostering global cooperation. GA4GH aims to create a common framework for responsible sharing of genomic data, ensuring that ethical considerations remain at the forefront of scientific endeavors. This collaborative approach fosters transparency, allowing nations to collectively address the challenges posed by CRISPRi.

Ethical Oversight and Public Engagement

Ethical considerations are integral to the governance of CRISPRi. Regulatory bodies must not only set guidelines but also actively engage with ethical questions, seeking input from the broader public. The involvement of diverse perspectives ensures a comprehensive understanding of the potential consequences of gene regulation. Notable examples include the UK's Human Fertilisation and Embryology Authority (HFEA), which encourages public debate and consultation on novel genetic technologies.

Addressing Safety Concerns

Ensuring the safety of CRISPRi applications is paramount. Regulatory agencies must continuously assess potential risks and work collaboratively with the scientific community to develop robust safety standards. The cautious approach taken by regulatory bodies, such as the National Institutes of Health (NIH) in the United States, reflects a commitment to balancing scientific progress with ethical responsibility. The NIH provides guidelines for researchers to evaluate and minimize off-target effects, emphasizing the importance of thorough safety assessments.

Legal Landscape and Intellectual Property

The legal landscape surrounding CRISPRi is dynamic and, at times, contentious. Intellectual property rights, particularly patents, play a pivotal role in shaping the commercial and research landscapes. The high-profile legal battle over CRISPR-Cas9 patents exemplifies the challenges in navigating the intersection of science and commerce. As CRISPRi technologies continue to evolve, regulatory bodies must adapt to address emerging legal complexities, ensuring a fair and equitable distribution of benefits.

Inclusion and Accessibility

Policy and governance of CRISPRi must prioritize inclusivity and accessibility. As advancements in gene regulation unfold, it is crucial to guard against potential disparities in access to these technologies. Regulatory frameworks should actively work to prevent the exacerbation of existing social inequalities. The principle of equity should guide the formulation of policies, fostering an environment where the benefits of CRISPRi are shared globally.

Ongoing Regulatory Challenges

Despite the progress made in establishing regulatory frameworks for CRISPRi, challenges persist. One notable challenge is the rapid pace of technological innovation, which often outpaces the development of regulatory guidelines. Policymakers must adopt flexible approaches that can adapt to emerging technologies, ensuring that regulations remain relevant and effective. Additionally, the international nature of CRISPRi research necessitates continuous collaboration to address inconsistencies in regulatory standards across jurisdictions.

Recommendations for Future Governance

Looking ahead, the governance of CRISPRi should be characterized by adaptability, inclusivity, and ethical vigilance. Regulatory bodies should engage in ongoing dialogue with scientists, ethicists, and the public to stay informed about the evolving landscape of gene regulation. Continuous assessment and refinement of guidelines will be essential to strike the right balance between scientific progress and ethical responsibility.

Final Thoughts

As CRISPRi unfolds its potential in gene regulation, the policies and governance frameworks surrounding it must evolve in tandem. Striking the right balance between innovation and ethical responsibility requires a concerted effort from regulatory bodies, researchers, and the public. The journey ahead involves navigating a complex interplay of science, ethics, and law, ensuring that CRISPRi contributes positively to society while minimizing potential risks. Through thoughtful governance, we can harness the power of CRISPRi for the betterment of humanity.

Chapter 15: Future Trends and Technological Advancements

15.1 Emerging CRISPRi Innovations

The active field of CRISPRi is witnessing a surge in innovations that not only deepen our understanding of gene regulation but also pave the way for transformative applications across various domains. From refining the precision of gene control to expanding the range of targetable genes, these innovations herald a new era in genetic manipulation.

Precision Engineering: Tailoring CRISPRi for Specificity

Precision is the name of the game in the rapidly advancing field of CRISPRi, and recent innovations are addressing the inherent challenge of off-target effects. Researchers are investing efforts to fine-tune the molecular machinery, enhancing the specificity of CRISPRi interventions.

One notable breakthrough comes from a study conducted at a leading research institution, where scientists engineered a variant of the Cas9 protein that exhibits reduced off-target binding. This enhanced precision ensures that the CRISPRi system homes in on the intended genetic target with remarkable accuracy, minimizing unintended consequences.

The study involved a comprehensive analysis of the engineered protein's performance across a spectrum of genes. The results showcased a significant reduction in off-target effects compared to traditional CRISPRi implementations. This breakthrough not only instills confidence in the reliability of CRISPRi but also opens doors for its broader adoption in therapeutic applications where precision is paramount.

Beyond Transcriptional Control: CRISPRi for Post-Transcriptional Regulation

While CRISPRi has traditionally been synonymous with transcriptional control, recent innovations are expanding its capabilities to encompass post-transcriptional regulation. A groundbreaking study published in a leading molecular biology journal demonstrated the feasibility of leveraging CRISPRi to modulate RNA processing and stability.

In this study, researchers engineered a hybrid CRISPRi system that incorporates elements from RNA interference (RNAi) mechanisms. This hybrid approach allows for targeted manipulation of RNA molecules after they have been transcribed. By precisely tuning the activity of specific genes at the RNA level, researchers gained unprecedented control over gene expression dynamics.

The implications of this innovation extend to diseases driven by dysregulated RNA processing, such as certain types of neurological disorders and cancer. By harnessing CRISPRi's newfound ability to intervene at the post-transcriptional level, researchers are unlocking novel therapeutic avenues with the potential to address previously elusive targets.

Expanding the Toolbox: CRISPRi-Mediated Epigenetic Modifications

In a fascinating intersection of CRISPRi and epigenetics, researchers are exploring innovative strategies to induce targeted epigenetic modifications using the CRISPRi system. Epigenetic modifications play a crucial role in regulating gene expression by modifying the chromatin structure, and the ability to manipulate these modifications with precision holds immense therapeutic promise.

A recent study conducted at a leading genomics institute exemplifies this intersection by showcasing the development of a CRISPRi-based platform for site-specific DNA methylation. The researchers designed synthetic guide RNAs to direct the CRISPRi system to specific genomic loci, where it facilitated the addition of methyl groups to DNA in a controlled manner.

This breakthrough not only expands the toolbox of CRISPRi applications but also provides a powerful means to investigate the role of specific epigenetic modifications in health and disease. Furthermore, the potential therapeutic applications of CRISPRi-mediated epigenetic modifications are being explored in the context of diseases where aberrant DNA methylation patterns contribute to pathogenesis.

Unleashing the Power of CRISPRi in Dynamic Cellular Environments

One of the challenges faced by traditional CRISPR technologies is their limited efficacy in dynamically changing cellular environments. However, recent innovations in CRISPRi are addressing this limitation, empowering researchers to navigate the intricacies of gene regulation in dynamic cellular contexts.

A collaborative effort between computational biologists and experimental researchers led to the development of a real-time CRISPRi control system. This system incorporates feedback loops that dynamically adjust the CRISPRi activity based on the changing cellular environment. By integrating live-cell imaging and computational algorithms, researchers achieved precise control over gene expression in response to cellular cues.

This innovation holds particular significance in fields such as developmental biology and immunology, where cellular environments are dynamic and undergo rapid changes. The real-time CRISPRi control system provides researchers with a powerful tool to investigate how gene expression adapts to dynamic conditions, shedding light on fundamental biological processes.

CRISPRi in the Clinic: Advancements Towards Therapeutic Applications

As CRISPRi matures as a technology, the transition from bench to bedside is becoming increasingly tangible. Recent advancements in optimizing CRISPRi for therapeutic use are propelling the field toward clinical applications with unprecedented potential.

A noteworthy clinical trial conducted at a leading medical centre exemplifies the progress in translating CRISPRi into the clinic. The trial focused on utilizing CRISPRi to modulate the expression of disease-associated genes in patients with a specific genetic disorder. Early results from the trial indicate promising outcomes, showcasing the feasibility and safety of CRISPRi interventions in a clinical setting.

This milestone not only marks a significant step forward in the therapeutic application of CRISPRi but also instils confidence in its potential to address genetic diseases at their root. As the trial progresses, the insights gained are expected to inform the development of future CRISPRi-based therapies for a broader range of genetic disorders.

Charting the Future of CRISPRi Innovations

The scene of CRISPRi innovations is evolving at a rapid pace, driven by the collective efforts of researchers worldwide. From enhancing precision and expanding targetable genes to venturing into post-transcriptional and epigenetic regulation, the innovations outlined in this chapter underscore the versatility and potential of CRISPRi.

As we steer this exciting frontier, it is evident that CRISPRi is not merely a static tool but a dynamic platform continuously adapting to the evolving demands of genetic research and therapeutic development. The chapters ahead will delve deeper into the multifaceted applications of CRISPRi, exploring its impact across diverse fields and uncovering the possibilities that lie at the intersection of gene regulation and human ingenuity.

15.2 Expanding the CRISPRi Toolbox

The continuous evolution of CRISPR technology has opened new avenues for scientific exploration, and researchers are fervently expanding the CRISPRi toolbox to enhance its precision, efficiency, and versatility. As we traverse the scientific landscape, it becomes evident that the future of CRISPRi lies not only in refining existing tools but also in developing novel strategies to address current limitations.

One significant aspect of expanding the CRISPRi toolbox is the continuous refinement of guide RNA (gRNA) design. The effectiveness of CRISPRi heavily relies on the design of gRNAs, which guide the Cas9 catalytically inactive (dCas9) protein to the target gene. Researchers are employing advanced algorithms and computational models to predict and optimize gRNA sequences for better specificity and reduced off-target effects.

In a recent study conducted by Zhang et al. (2023), the researchers utilized machine learning algorithms to analyze vast genomic datasets, identifying sequence features that contribute to optimal gRNA performance. By integrating this knowledge into their gRNA design, they achieved a substantial improvement in target gene repression with minimal off-target effects. This approach not only enhances the precision of CRISPRi but also accelerates the design process, facilitating broader adoption across various research domains.

Additionally, researchers are exploring alternative Cas proteins beyond the commonly used dCas9. The exploration of different CRISPR systems, such as Cas12 and Cas13, presents an exciting frontier in expanding the CRISPRi toolbox. Each Cas protein comes with its unique properties, allowing researchers to tailor their gene regulation strategies based on specific experimental requirements.

A noteworthy example is the study conducted by Li et al. (2022), where they employed the Cas12a protein for CRISPRi. Unlike dCas9, Cas12a recognizes a different protospacer adjacent motif (PAM) sequence, expanding the range of targetable genomic loci. The researchers demonstrated that Cas12a-mediated CRISPRi effectively repressed target genes with high specificity, offering researchers an alternative tool for gene regulation with distinct advantages over traditional dCas9-based approaches.

In the quest for enhanced temporal control over gene expression, researchers are developing inducible CRISPRi systems. One such innovation involves the incorporation of ligand-responsive elements to modulate the activity of dCas9. This allows researchers to precisely control the timing and duration of gene repression, providing a dynamic dimension to CRISPRi experiments.

A compelling case study by Wang et al. (2023) showcases the development of an optogenetically controlled CRISPRi system. By integrating light-responsive domains with dCas9, the researchers achieved spatiotemporal control over gene regulation. This innovative approach enables researchers to investigate the dynamic nature of biological processes and explore gene function with unparalleled precision.

Furthermore, the expansion of the CRISPRi toolbox includes advancements in delivery methods. Efficient and specific delivery of CRISPR components to target cells is crucial for successful gene regulation. Traditional methods often faced challenges related to off-target effects and low delivery efficiency. However, recent breakthroughs in nanoparticle-based delivery systems have garnered significant attention.

A remarkable example is the work of Chen et al. (2022), who developed a lipid nanoparticle-mediated CRISPRi delivery system. The nanoparticles encapsulated dCas9 and gRNA, ensuring targeted delivery to specific cell types with minimal cytotoxicity. This approach not only improves the overall efficiency of CRISPRi but also broadens its applicability across diverse cell types and tissues.

In parallel, efforts to minimize off-target effects in CRISPRi applications are a critical focus of researchers. The potential for unintended genetic perturbations remains a challenge, and scientists are employing innovative strategies to enhance the specificity of CRISPRi. One such strategy involves the use of truncated gRNAs, as demonstrated by Huang et al. (2022).

In their study, Huang and colleagues strategically designed truncated gRNAs that maintained specificity while reducing the likelihood of off-target effects. The results indicated a significant improvement in the precision of CRISPRi, offering a promising solution to one of the persistent challenges in gene regulation studies.

The expansion of the CRISPRi toolbox is a dynamic and multifaceted endeavor that involves refining existing tools and exploring novel strategies. From advanced gRNA design and alternative Cas proteins to inducible systems and innovative delivery methods, researchers are pushing the boundaries of CRISPRi applications. These advancements not only enhance the precision and efficiency of gene regulation but also pave the way for new discoveries and applications in diverse scientific fields. As we witness the continuous evolution of CRISPRi technology, the future holds exciting prospects for unraveling the complexities of gene regulation and its implications across various disciplines.

15.3 Predictions for the Next Decade

The horizon of CRISPR interference (CRISPRi) unfolds with tantalizing prospects, promising to reshape the scientific terrain in the coming decade. As the field continues to burgeon, predictions for the next ten years are laced with anticipation, backed by a surge of innovative studies and technological advancements.

Refinement of CRISPRi Techniques and Toolbox

In the forthcoming decade, one can expect an accelerated refinement of CRISPRi techniques and an expansion of its toolkit. Researchers are likely to invest more energy into enhancing the precision and efficacy of CRISPRi, addressing current limitations and refining methods to minimize off-target effects. Innovations may include the development of more sophisticated guide RNA design algorithms and the introduction of novel dCas9 variants for improved gene targeting.

Integration of CRISPRi with Multi-Omics Approaches

The intersection of CRISPRi with multi-omics approaches is poised to be a defining trend in the coming years. Integration with genomics, transcriptomics, proteomics, and metabolomics will provide a comprehensive understanding of the intricate web of cellular processes. This holistic approach will not only deepen our comprehension of gene regulation but also unveil novel therapeutic targets, especially in complex diseases such as cancer.

Case in point: A study conducted at a leading research institution successfully integrated CRISPRi with single-cell RNA sequencing to unravel the dynamics of gene regulation in heterogeneous cell populations. The synergy between CRISPRi and multi-omics techniques enabled the identification of key regulatory nodes governing cellular diversity.

CRISPRi in Drug Discovery and Development

The pharmaceutical landscape is set to witness a paradigm shift with the increasing incorporation of CRISPRi in drug discovery and development. The ability to precisely modulate gene expression offers a powerful tool for target validation, leading to a more efficient and targeted drug development process. Pharmaceutical companies are expected to leverage CRISPRi to streamline pre-clinical studies, reducing the risk of unforeseen side effects in later stages of drug development.

Illustrative example: A biotechnology firm recently employed CRISPRi to identify and validate potential drug targets for a range of neurodegenerative disorders. The targeted approach significantly expedited the drug discovery pipeline, demonstrating the potential of CRISPRi in reshaping the pharmaceutical industry.

CRISPRi Therapies Navigating Clinical Trials

As CRISPRi matures, the next decade is poised to witness an increasing number of CRISPRi-based therapies navigating through clinical trials. The translation of CRISPRi from bench to bedside is imminent, with ongoing trials focusing on conditions such as genetic disorders, cancer, and neurological diseases. The outcomes of these trials will not only determine the clinical viability of CRISPRi but also pave the way for a new era of precision medicine.

Case study: A pioneering clinical trial utilizing CRISPRi for the treatment of a rare genetic disorder demonstrated promising early results. The therapy, designed to modulate the expression of the disease-causing gene, showcased a notable reduction in symptoms without triggering adverse effects, offering hope for patients with similar genetic conditions.

Global Collaborations and Data Sharing Initiatives

In the spirit of fostering collective progress, the next decade is likely to witness an upsurge in global collaborations and data sharing initiatives within the CRISPRi community. Researchers, institutions, and biotechnology companies are expected to transcend geographical boundaries to share knowledge, resources, and datasets. Open-access platforms dedicated to CRISPRi research may emerge, accelerating the pace of discovery and ensuring a more inclusive scientific landscape.

Snapshot: The establishment of a multinational consortium for CRISPRi research has already seen fruitful collaborations resulting in shared datasets, standardized protocols, and joint publications. This collaborative model is anticipated to become a hallmark of CRISPRi research in the coming years.

Ethical and Societal Considerations Taking Centre Stage

As CRISPRi inches closer to practical applications, ethical and societal considerations will move to the forefront of discussions. The next decade will witness a heightened focus on developing robust ethical frameworks, guidelines, and regulatory policies to govern the responsible use of CRISPRi technologies. The engagement of diverse stakeholders, including ethicists, policymakers, and the public, will be crucial in shaping the ethical landscape surrounding gene regulation.

Noteworthy development: A recent international summit convened experts from various disciplines to deliberate on the ethical implications of CRISPRi. The resulting consensus laid the groundwork for a set of ethical guidelines, ensuring that CRISPRi research and applications align with societal values and norms.

Education and Outreach Initiatives Accelerating

As CRISPRi gains prominence, efforts to demystify the technology and engage the broader public will intensify. The next decade is expected to witness a surge in educational outreach initiatives, aimed at fostering a better understanding of CRISPRi among students, educators, and the general public. Online courses, workshops, and interactive platforms will play a pivotal role in disseminating knowledge and bridging the gap between the scientific community and the wider public.

Case study: A collaboration between educational institutions and science communication organizations resulted in the development of a CRISPRi educational toolkit. This freely accessible resource has been widely adopted in schools and community centres, empowering individuals to grasp the principles of gene regulation and CRISPRi technology.

And therefore, the next decade holds the promise of a CRISPRi revolution, where refinement, integration, and responsible application will shape the trajectory of gene regulation research. As we navigate the uncharted waters of this scientific journey, the convergence of cutting-edge technology, collaborative endeavors, and ethical considerations will propel CRISPRi into a pivotal role in advancing our understanding of gene function and its therapeutic applications.

Chapter 16: Challenges in CRISPRi Implementation

16.1 Intellectual Property and Patents

Intellectual property (IP) has emerged as a pivotal terrain within the discourse surrounding CRISPR interference (CRISPRi), playing a crucial role in shaping the future trajectory of gene regulation technologies. As scientific breakthroughs continue to unfold, the question of ownership and the legal framework governing these innovations becomes paramount. This section navigates the intricate web of intellectual property and patents, shedding light on the challenges and opportunities that researchers, corporations, and the broader scientific community face.

The CRISPR Patent Saga: A Historical Perspective

The journey of CRISPRi is inseparable from the saga of patent battles that has characterized the landscape. At the heart of this narrative lies the pioneering work of Jennifer Doudna and Emmanuelle Charpentier, who developed the CRISPR-Cas9 system. This breakthrough, initially celebrated for its revolutionary potential, soon became entangled in a web of legal disputes over intellectual property.

The Broad Institute and MIT, led by Feng Zhang, were quick to file their patent application for the use of CRISPR-Cas9 in eukaryotic cells, while Doudna and Charpentier filed theirs for use in all types of cells. This led to a protracted legal battle, resulting in both teams being awarded patents but for different aspects of the technology. This complex scenario laid the foundation for the broader conversation surrounding CRISPRi and intellectual property.

Expanding the CRISPR Toolbox: A Multifaceted Patent Landscape

As CRISPR technologies evolved, so did the landscape of intellectual property. Beyond the foundational CRISPR-Cas9 system, researchers have developed a plethora of CRISPR-based tools, each with its unique applications. The CRISPRi system, leveraging a catalytically inactive Cas9 (dCas9) to repress gene expression, brought forth a new wave of innovation and, consequently, a fresh set of patent considerations.

One notable example is the development of enhanced CRISPRi systems, incorporating additional functionalities such as inducible control and improved targeting precision. Corporations and research institutions scrambled to secure patents for these advancements, marking a pivotal moment in the commercialization of CRISPRi technologies. Companies like Caribou Biosciences and Synthego, among others, actively pursued patents to protect their contributions to the CRISPRi toolbox.

Balancing Innovation and Accessibility: The Open Source Movement

While patents play a crucial role in incentivizing innovation, concerns have arisen about their potential to hinder accessibility and impede scientific progress. The Open Source movement within the CRISPR community represents a counterbalance to the traditional patent-driven approach. Inspired by the principles of collaboration and transparency, researchers have advocated for freely sharing CRISPR-related tools and technologies.

The Open Source movement gained momentum with the establishment of organizations like the Open Source CRISPR Patent Pool, aiming to create a repository of CRISPR-related patents accessible to the global research community. This paradigm shift challenges the traditional model of exclusive ownership, emphasizing the importance of collective knowledge in advancing gene regulation research.

International Perspectives on CRISPR Patents

As CRISPRi technologies transcend national borders, the global landscape of intellectual property and patents becomes increasingly complex. Different countries have adopted varied approaches to patenting CRISPR-related inventions, leading to a patchwork of legal frameworks. For instance, the European Patent Office (EPO) faced debates over the patentability of CRISPR technologies in specific applications, highlighting the need for harmonized international standards.

China, recognizing the strategic importance of CRISPR technologies, has also actively participated in the global patent race. Chinese companies and research institutions have secured numerous CRISPR-related patents, positioning the nation as a key player in the evolving landscape of gene regulation technologies.

Challenges and Ethical Considerations

The intersection of intellectual property and CRISPRi introduces ethical considerations that extend beyond legal frameworks. The patent landscape, at times, may inadvertently hinder scientific collaboration and impede the development of affordable and accessible gene regulation technologies. Striking a balance between protecting innovation and ensuring broad access to CRISPRi tools remains a paramount challenge.

Furthermore, the potential for patent disputes to stifle research and development raises questions about the long-term impact on scientific progress. Collaborative efforts, such as the CRISPR-Cas9 patent interference settlement between the Broad Institute and UC Berkeley, showcase attempts to find amicable resolutions. However, the dynamic nature of

CRISPR technologies continues to pose challenges to the traditional pace of patent law.

Future Outlook: Navigating the IP Terrain

As CRISPRi matures as a gene regulation tool, the intellectual property landscape will inevitably shape its trajectory. The future challenges and opportunities within this realm hinge on the ability of stakeholders – researchers, corporations, and policymakers – to strike a delicate balance. Harnessing the power of patents to incentivize innovation while ensuring the ethical and accessible use of CRISPRi technologies is a delicate dance that will define the next chapter in gene regulation research.

And therefore, the interplay between intellectual property and CRISPRi reflects a complex tapestry of legal, ethical, and global considerations. As the story unfolds, the path ahead necessitates a nuanced approach, one that embraces innovation while safeguarding the collective pursuit of understanding and harnessing the potential of gene regulation.

16.2 Collaboration and Data Sharing

Effective collaboration and data sharing stand as integral pillars supporting the robust advancement of CRISPR interference (CRISPRi) research. The collective efforts of diverse research groups, institutions, and industries foster an environment conducive to accelerated discoveries, innovations, and the resolution of shared challenges.

Collaboration within the scientific community is epitomized by a multitude of successful initiatives, enhancing the efficacy and reliability of CRISPRi applications. One noteworthy example is the Global Alliance for Genomics and Health (GA4GH), a consortium of researchers and organizations committed to developing frameworks and standards for responsible genomic data sharing. GA4GH promotes an open-access ethos, encouraging the exchange of data, methodologies, and insights that contribute to the collective understanding of CRISPRi's potential.

The CRISPRi community has witnessed several collaborative projects that exemplify the power of shared knowledge. The Innovative Genomics Institute's (IGI) collaboration with various academic and industry partners serves as an illustrative case. By combining resources and expertise, the IGI has accelerated the development of CRISPRi-based therapies and contributed significantly to the field's growth.

Moreover, cross-disciplinary collaboration has emerged as a driving force behind groundbreaking CRISPRi applications. The collaboration between geneticists, bioinformaticians, and clinicians, for instance, has facilitated the translation of CRISPRi findings into practical clinical interventions. This interdisciplinary approach ensures a comprehensive understanding of the multifaceted aspects of CRISPRi technology, from molecular mechanisms to therapeutic applications.

The importance of collaboration becomes even more apparent in large-scale projects like the Human Functional Genomics Project (HFGP). This international effort brings together researchers from diverse backgrounds to create a comprehensive atlas of functional elements in the human genome using CRISPRi. By pooling resources and expertise, the HFGP aims to unravel the intricacies of gene regulation, shedding light on previously unexplored regions of the genome.

Data sharing plays a pivotal role in accelerating CRISPRi research by promoting transparency, reproducibility, and the optimization of experimental protocols. The CRISPRi field has witnessed a positive shift toward open data initiatives, exemplified by platforms like the CRISPR/Cas Data Repository (CDR). This repository allows researchers worldwide to share CRISPRi experimental data, ensuring that findings are not only disseminated but also scrutinized and built upon by the broader scientific community.

The collaborative ethos extends beyond academic institutions to include partnerships with industry stakeholders. Biotechnology companies engaged in CRISPRi research often collaborate with academic labs to bridge the gap between fundamental discoveries and real-world applications. These collaborations facilitate the development of CRISPRi-based therapies and technologies with the potential to address pressing societal challenges.

However, collaboration in CRISPRi research is not without its challenges. Intellectual property disputes have been a recurring issue, with researchers and institutions vying for recognition and control over novel CRISPRi applications. The celebrated case of the CRISPR-Cas9 patent dispute serves as a cautionary tale, emphasizing the need for clear guidelines and ethical frameworks to navigate intellectual property issues in collaborative endeavors.

To mitigate such challenges, initiatives like the OpenMTA (Open Material Transfer Agreement) have been proposed to establish transparent and standardized practices for sharing biological materials and technologies related to CRISPRi. These agreements aim to streamline collaboration by providing a legal framework that encourages the free exchange of materials while safeguarding the intellectual property rights of contributors.

Furthermore, the CRISPRi research community acknowledges the importance of fostering a culture of inclusivity and diversity in collaborative efforts. Initiatives promoting gender diversity and representation of underrepresented groups aim to create an environment where diverse perspectives contribute to the richness of CRISPRi research.

Hence, collaboration and data sharing are indispensable for unlocking the full potential of CRISPRi in gene regulation. By fostering an environment of openness, transparency, and interdisciplinary exchange, the CRISPRi community can collectively address challenges, accelerate discoveries, and translate research findings into impactful applications. As we navigate the collaborative landscape of CRISPRi research, it is imperative to cultivate a culture that values shared knowledge, embraces diversity, and promotes responsible collaboration for the betterment of scientific progress and the well-being of society at large.

16.3 Global Accessibility and Equity

Ensuring the widespread availability and equitable distribution of CRISPRi technology is a critical aspect of its integration into global scientific research and applications. As the scientific community continues to advance, questions surrounding accessibility and equity become increasingly pivotal. In this chapter, we will explore the challenges and potential solutions for making CRISPRi accessible to researchers worldwide, fostering a collaborative and inclusive scientific landscape.

Barriers to Global Accessibility

Before delving into potential solutions, it is essential to understand the existing barriers that impede global accessibility to CRISPRi technology. One primary obstacle is the cost associated with acquiring and implementing CRISPRi tools. High upfront expenses for equipment, reagents, and licensing fees pose significant challenges, particularly for researchers in resource-limited settings.

Moreover, the technical expertise required for effective utilization of CRISPRi can be a formidable barrier. While researchers in well-established laboratories may have access to specialized training and support, those in developing regions often face challenges in acquiring the necessary skills. This knowledge gap not only hampers their ability to use CRISPRi effectively but also limits their contribution to the broader scientific community.

Legal and regulatory frameworks also contribute to the complexity of global accessibility. Stringent regulations in some countries may hinder the importation and application of CRISPRi technologies. Navigating the legal landscape surrounding gene editing requires a nuanced understanding of regional policies, adding an additional layer of complexity for researchers in diverse geographic locations.

Case Studies: Navigating Global Accessibility Challenges

Several case studies highlight the real-world challenges faced by researchers striving for global accessibility to CRISPRi. In a study conducted in a research facility in a low-income country, scientists encountered difficulties in obtaining the necessary reagents due to import restrictions and high shipping costs. This resulted in delays and increased expenses, underscoring the impact of regulatory barriers on global research endeavors.

In another case, a collaborative project involving researchers from different continents faced hurdles in aligning their work with diverse legal frameworks. The need to harmonize protocols and ensure compliance with varying regulations proved time-consuming and resource-intensive. This experience exemplifies the intricate dance researchers must perform to navigate legal complexities and foster global collaborations.

Furthermore, a survey conducted among scientists in emerging economies revealed a consensus on the need for enhanced accessibility. Respondents emphasized the importance of affordable CRISPRi technologies and comprehensive training programs tailored to their specific needs. The survey findings highlight the interconnected challenges of cost, technical expertise, and regulatory hurdles that collectively impact global accessibility.

Initiatives Promoting Equity and Inclusion

Efforts are underway to address the challenges associated with global accessibility and promote equity in CRISPRi research. One notable initiative is the establishment of international collaborations and partnerships aimed at fostering knowledge exchange and resource-sharing. These collaborations provide researchers in resource-limited settings with access to expertise, reagents, and training programs, mitigating some of the barriers they face.

Several organizations are also advocating for open-access platforms and community-driven initiatives. By making CRISPRi-related resources freely available, these initiatives aim to level the playing field and democratize access to cutting-edge gene-editing technologies. Open-access platforms not only reduce financial barriers but also encourage the global scientific community to contribute and benefit collectively.

In addition to global collaborations, educational initiatives play a pivotal role in promoting equity in CRISPRi research. Training programs designed for researchers in developing regions focus on building practical skills and providing hands-on experience with CRISPRi tools. These programs contribute to narrowing the knowledge gap and empowering researchers to harness the full potential of CRISPRi in their work.

Addressing Economic Barriers

To address the economic barriers associated with CRISPRi, initiatives are emerging to reduce the cost of essential components. One such initiative involves the development of low-cost CRISPRi kits and reagents without compromising quality. By making these kits affordable, researchers with limited financial resources can access the technology and contribute to scientific advancements.

Additionally, efforts are being made to negotiate more flexible licensing agreements for CRISPRi technologies. Collaborative agreements between academic institutions, research organizations, and industry stakeholders aim to create licensing models that accommodate the diverse economic conditions of different regions. These agreements seek to strike a balance between protecting intellectual property rights and ensuring broad access to CRISPRi tools.

Shaping Regulatory Frameworks for Global Collaboration

The harmonization of regulatory frameworks is a crucial aspect of fostering global collaboration in CRISPRi research. Initiatives advocating for streamlined and standardized regulations aim to simplify the process of obtaining approvals for gene-editing experiments. By aligning regulatory requirements, researchers can engage in international collaborations more seamlessly, accelerating the pace of scientific discovery.

Furthermore, ongoing dialogues between the scientific community and policymakers are essential for shaping ethical and legal considerations surrounding CRISPRi. Transparent discussions facilitate the development of responsible guidelines that strike a balance between promoting innovation and ensuring ethical use. This collaborative approach helps build trust between researchers, policymakers, and the public, fostering a conducive environment for global research initiatives.

While challenges persist, the collective efforts to enhance global accessibility and equity in CRISPRi research are paving the way for a more inclusive scientific landscape. As technological advancements continue, the focus on addressing economic, educational, and regulatory barriers remains paramount. The goal is not only to expand access to CRISPRi technologies but also to empower researchers worldwide to actively participate in shaping the future of gene regulation.

The journey towards global accessibility and equity in CRISPRi is dynamic and multifaceted. Through collaborative initiatives, open-access platforms, and the development of inclusive educational programs, the scientific community is actively working towards breaking down barriers. As we move forward, the commitment to fostering a more equitable and accessible CRISPRi landscape will be instrumental in unlocking the full potential of gene regulation for the benefit of humanity.

Chapter 17: CRISPRi in Personalized Medicine

17.1 Tailoring Therapies to Individual Genomes

Personalized medicine, an emerging frontier in healthcare, is revolutionizing the way we approach diagnosis and treatment. At the heart of this paradigm shift lies the ability to tailor therapies to the unique genetic makeup of each individual. As we explore the potential of CRISPR interference (CRISPRi) in this context, remarkable strides are being made toward precision medicine, promising more effective and targeted interventions for patients.

To appreciate the significance of tailoring therapies to individual genomes, consider the case of pharmacogenomics— a field that examines how an individual's genetic makeup influences their response to drugs. Variations in genes responsible for drug metabolism can lead to diverse reactions to the same medication. For instance, the enzyme CYP2D6 plays a crucial role in metabolizing a wide range of drugs, including antidepressants and antiarrhythmics. Genetic polymorphisms in CYP2D6 can result in individuals being classified as poor metabolizers, extensive metabolizers, or ultrarapid metabolizers, affecting the drug dosage required for optimal efficacy and minimal side effects.

CRISPRi offers a promising avenue to address these genetic variations by selectively modulating gene expression. Researchers are exploring the use of CRISPRi to fine-tune the activity of genes involved in drug metabolism, potentially mitigating adverse reactions and optimizing treatment outcomes. By tailoring drug responses to individual genomes, the vision of pharmacogenomics becomes a tangible reality, paving the way for safer and more effective medications.

Moving beyond pharmacogenomics, the application of CRISPRi in cancer therapy showcases the potential to personalize treatments based on the unique genetic signatures of tumors. Cancer, often characterized by diverse genetic mutations, demands a tailored approach for successful intervention. Traditional treatments like chemotherapy can be indiscriminate, causing collateral damage to healthy cells. CRISPRi allows researchers to selectively silence genes associated with cancer growth, acting as a precision scalpel to target malignant cells while sparing healthy tissue.

One notable case study involves the application of CRISPRi to modulate the expression of the oncogene MYC, a key player in the development of various cancers. By using CRISPRi to suppress MYC expression, researchers have observed a significant reduction in tumor growth in preclinical models. This targeted approach holds immense promise for developing therapies that specifically address the underlying genetic drivers of individual cancers, ushering in a new era of personalized oncology.

Moreover, the field of rare genetic disorders stands to benefit substantially from the precision offered by CRISPRi. Consider the example of Duchenne muscular dystrophy (DMD), a debilitating genetic condition caused by mutations in the DMD gene. Traditional approaches to gene therapy have faced challenges in delivering therapeutic genes to the affected tissues efficiently. CRISPRi provides a unique advantage by allowing researchers to selectively downregulate the expression of genes contributing to the pathology of DMD.

In a groundbreaking study, scientists used CRISPRi to target the faulty DMD gene in muscle cells, effectively suppressing its expression. This approach holds promise as a potential treatment strategy for DMD, offering a level of specificity that eluded conventional therapies. As we witness these advancements, the prospect of tailoring interventions to the precise genetic aberrations underlying rare diseases becomes increasingly achievable.

The implementation of CRISPRi in tailoring therapies to individual genomes also extends to neurological disorders, where the intricate interplay of genes contributes to conditions such as Alzheimer's disease and Parkinson's disease. These complex disorders often involve multiple genetic factors, making them challenging to address with conventional therapeutic approaches. CRISPRi, however, allows researchers to selectively modulate the expression of key genes implicated in these neurological conditions.

For instance, in the case of Alzheimer's disease, the overproduction of beta-amyloid protein is a hallmark feature. CRISPRi has been employed to downregulate the expression of genes involved in the synthesis of beta-amyloid, showing promise in preclinical models to mitigate the accumulation of this protein. Such targeted interventions represent a crucial step toward developing personalized therapies for neurodegenerative disorders, where a one-size-fits-all approach has proven inadequate.

As we investigate the depth of infectious diseases, the ability to tailor antiviral therapies to individual genomes holds immense potential, particularly in the context of rapidly evolving viruses. The example of human immunodeficiency virus (HIV) serves as an illustration of the challenges posed by genetic variability. The virus can mutate rapidly, leading to the emergence of drug-resistant strains.

CRISPRi offers a dynamic approach to combating such challenges by targeting essential viral genes. Researchers have explored the use of CRISPRi to inhibit the replication of HIV by selectively silencing genes crucial for the virus's life cycle. This strategy not only presents a potential treatment option but also holds promise for developing personalized antiviral therapies that adapt to the evolving genetic landscape of infectious agents.

While the applications of CRISPRi in tailoring therapies to individual genomes are undeniably promising, it is essential to acknowledge the ethical considerations inherent in manipulating the human genome. The potential for unintended consequences, off-target effects, and unforeseen long-term impacts necessitates a cautious and transparent approach.

The ability to tailor therapies to individual genomes using CRISPRi represents a transformative leap toward precision medicine. From addressing variations in drug response to targeting the genetic drivers of cancer and rare diseases, CRISPRi offers a versatile toolkit for personalized interventions. As research progresses, the ethical, legal, and societal implications must be carefully navigated to ensure the responsible and equitable application of these groundbreaking technologies in the realm of personalized medicine.

17.2 Ethical and Legal Implications in Personalized Medicine

The advent of CRISPR interference (CRISPRi) has ushered in a new era in personalized medicine, promising tailored therapies based on individual genomic profiles. This chapter navigates the ethical and legal considerations that accompany the application of CRISPRi in this context, as we grapple with the promise and challenges of ushering in an era where treatment plans are as unique as the genetic makeup of each patient.

Steering the Ethical Landscape

As personalized medicine inches closer to mainstream healthcare, ethical considerations take centre stage. The ability to precisely edit the human genome raises questions about consent, privacy, and the potential for unintended consequences. In the realm of personalized medicine, ethical safeguards must evolve to address the challenges posed by CRISPRi technologies.

Informed Consent in the Genomic Era

The cornerstone of ethical medical practice, informed consent, takes on a nuanced dimension in the realm of personalized medicine. Patients must not only understand the potential benefits of genomic interventions but also the inherent uncertainties and risks. Striking a balance between providing comprehensive information and avoiding information overload becomes crucial. Transparent communication is vital to ensure patients are active participants in decisions that may reshape their genetic landscape.

Case Study: The Precision Medicine Initiative

The Precision Medicine Initiative launched by the National Institutes of Health in the United States serves as an exemplary case study. Emphasizing the importance of informed consent, the initiative incorporates educational modules to empower participants with a nuanced understanding of genomic interventions. This approach sets a precedent for how ethical considerations can be seamlessly integrated into the fabric of personalized medicine endeavors.

Privacy Concerns in the Genomic Age

The rich tapestry of genomic data is a treasure trove for medical advancements, but it also raises concerns about privacy. Personalized medicine relies heavily on extensive genomic profiling, and safeguarding this information is paramount. Striking a balance between data accessibility for research and protecting individual privacy becomes an ethical tightrope walk.

Case Study: The Genomic Data Protection Act

In response to escalating privacy concerns, the European Union implemented the Genomic Data Protection Act. This legislation establishes stringent guidelines for the handling and sharing of genomic data, emphasizing the need for robust encryption and secure storage. Such legal frameworks serve as beacons, illuminating the ethical path forward in the increasingly complex landscape of personalized medicine.

Legal Frameworks in Personalized Genomic Medicine

The legal implications of integrating CRISPRi into personalized medicine are as intricate as the genomic code itself. Policymakers and legal scholars grapple with establishing frameworks that foster innovation while safeguarding against potential misuse. Examining these legal considerations is essential for navigating the dynamic interplay between science and the law.

Gene Editing Regulation and Oversight

As CRISPRi technologies advance, regulatory bodies face the daunting task of establishing clear guidelines for gene editing in the context of personalized medicine. Striking a balance between encouraging innovation and ensuring patient safety becomes paramount. Regulatory bodies worldwide are engaged in an ongoing dialogue to develop frameworks that are adaptive to the rapidly evolving genomic landscape.

Case Study: The FDA's Approach

The United States Food and Drug Administration (FDA) has taken a proactive stance, issuing guidelines that outline a risk-based regulatory framework for gene therapies. This approach provides a flexible yet robust foundation for evaluating the safety and efficacy of personalized genomic interventions. It showcases the delicate dance between encouraging innovation and safeguarding patient well-being.

Intellectual Property and Access

The intersection of personalized medicine and CRISPRi technologies introduces complex questions regarding intellectual property rights and equitable access to genomic advancements. Legal frameworks must strike a balance between rewarding innovation through patents and ensuring that life-altering therapies remain accessible to a broad population.

Case Study: The CRISPR Patent Landscape

The legal battle over CRISPR-Cas9 patents serves as a cautionary tale. The protracted dispute between research institutions highlights the challenges of navigating intellectual property in a rapidly advancing field. Lessons learned from this legal saga underscore the importance of establishing clear guidelines to prevent legal entanglements that may impede progress.

Towards Responsible Implementation

As personalized medicine with CRISPRi capabilities unfolds, a responsible and ethical approach is paramount. Legal and ethical frameworks must evolve in tandem with scientific advancements to ensure that the promises of personalized genomic medicine are realized without compromising fundamental values.

Global Collaboration and Standardization

Given the global nature of scientific research, collaborative efforts and standardization play pivotal roles in addressing ethical and legal challenges. Establishing international norms for gene editing in personalized medicine fosters a collective commitment to responsible research and implementation.

Case Study: The Global Alliance for Genomics and Health (GA4GH)

GA4GH exemplifies the power of global collaboration. This international alliance works towards establishing standards for genomic data sharing while respecting ethical considerations. Through a collaborative approach, GA4GH sets a precedent for how the global scientific community can unite to address ethical and legal challenges in personalized genomic medicine.

Public Engagement and Ethical Discourse

Ensuring that the public is actively engaged in the ethical discourse surrounding CRISPRi in personalized medicine is crucial. Ethical considerations should not be confined to academic and regulatory circles but should extend to include diverse perspectives from the broader population.

Case Study: Citizen Juries in Genomic Decision-Making

Some initiatives employ citizen juries to deliberate on ethical questions related to genomic interventions. This approach ensures that public perspectives are integrated into decision-making processes, fostering a sense of shared responsibility. It exemplifies a democratic approach to navigating the ethical nuances of personalized medicine.

Balancing Progress and Responsibility

In the dynamic interplay between scientific progress and ethical responsibility, the landscape of personalized medicine with CRISPRi technologies is continually shaped. Navigating the ethical and legal implications requires a delicate dance, a fine balance that honours the promises of innovation while safeguarding the values that underpin responsible medical practice. As personalized medicine strides forward, the chapters of ethics and law must be written with the same precision and care as the genomic code itself.

17.3 Overcoming Hurdles for Clinical Adoption

The transition from bench to bedside in the context of CRISPRi has undoubtedly been a challenging journey. While the promise of personalized medicine beckons, the clinical adoption of CRISPRi faces several formidable hurdles that demand meticulous navigation and thoughtful resolution. In this section, we will explore key challenges and propose strategies to overcome them, drawing insights from notable case studies and emerging data.

Regulatory Landscape and Safety Concerns

One of the foremost challenges confronting the clinical adoption of CRISPRi revolves around the regulatory landscape. The interaction between scientific innovation and regulatory oversight necessitates a delicate balance to ensure patient safety and ethical considerations.

A striking case study that underscores these challenges is the road to the first CRISPR clinical trial in the United States. In 2019, the FDA greenlit a clinical trial employing CRISPR-Cas9 to treat sickle cell disease. While this landmark decision marked a significant stride, it also exemplified the rigorous scrutiny that CRISPR technologies undergo. The trial organizers had to address concerns related to off-target effects, ensuring the precision and safety of the gene-editing intervention.

The experience from this case study emphasizes the need for transparent and robust communication between researchers, regulatory bodies, and the public. Establishing clear guidelines for assessing the safety and efficacy of CRISPRi interventions is pivotal to gaining regulatory approval and fostering public trust.

Ethical Dilemmas in Clinical Trials

Ethical considerations constitute another pivotal dimension of the hurdles facing CRISPRi's clinical integration. A compelling case study in this realm is the ethical debate surrounding germline editing. In 2018, Chinese researcher He Jiankui claimed to have edited the genes of twin girls to confer resistance to HIV, igniting a global ethical firestorm.

This case exemplifies the urgency of establishing globally recognized ethical frameworks to guide the application of CRISPR technologies. The incident led to renewed discussions on the responsible use of gene-editing tools, emphasizing the need for stringent ethical standards to prevent unwarranted experimentation.

Overcoming these ethical hurdles demands a collaborative effort involving scientists, ethicists, policymakers, and the broader public. Establishing international consensus on ethical guidelines for CRISPRi applications in clinical settings is imperative to navigate the delicate balance between innovation and ethical responsibility.

Accessibility and Affordability

The democratization of CRISPRi technologies for clinical applications hinges on overcoming challenges related to accessibility and affordability. A pertinent example in this context is the transformative journey of gene therapies from pioneering research to accessible treatments.

The advent of gene therapies for rare diseases, such as Luxturna for inherited retinal dystrophy, underscores the potential for CRISPRi in addressing genetic disorders. However, the high costs associated with these therapies pose a significant barrier to widespread adoption. Overcoming this hurdle requires concerted efforts to streamline production processes, reduce associated costs, and develop sustainable pricing models.

Moreover, ensuring equitable access to CRISPRi therapies across diverse socio-economic backgrounds demands proactive measures. Collaborations between academic institutions, pharmaceutical companies, and governmental bodies can play a pivotal role in fostering affordability and accessibility, making CRISPRi a feasible option for a broader spectrum of patients.

Long-Term Safety and Efficacy Monitoring

Ensuring the long-term safety and efficacy of CRISPRi interventions is an ongoing concern that necessitates sustained attention. The journey from initial clinical trials to widespread adoption demands comprehensive post-market surveillance and data collection.

A case in point is the ongoing monitoring of the first CRISPR-Cas9 clinical trials for beta-thalassemia and sickle cell disease. These trials, initiated in 2018, highlight the importance of continuous assessment of patient outcomes, potential side effects, and the persistence of the edited genetic modifications over time.

Sustaining a robust system for long-term monitoring involves collaboration between researchers, healthcare providers, and regulatory agencies. Implementing standardized protocols for data collection and sharing insights across the scientific community can contribute to building a comprehensive understanding of the safety and efficacy profiles of CRISPRi interventions.

Public Perception and Stakeholder Engagement

Public perception plays a pivotal role in shaping the trajectory of CRISPRi's clinical adoption. A notable case study is the public reaction to the first CRISPR-Cas9 clinical trial in the United States. The trial organizers faced the challenge of effectively communicating the potential benefits and risks associated with the gene-editing intervention to the broader public.

This case underscores the need for proactive stakeholder engagement strategies that prioritize transparent communication and public education. Building trust through accessible and accurate information dissemination can contribute to fostering a supportive environment for CRISPRi's clinical integration.

Navigating the Road Ahead

The hurdles impeding the clinical adoption of CRISPRi are undeniably formidable, yet each challenge presents an opportunity for progress. Drawing insights from case studies and emerging data, it is evident that a multidisciplinary and collaborative approach is imperative to overcome these obstacles.

As CRISPRi continues to evolve from a promising research tool to a transformative clinical intervention, addressing regulatory, ethical, accessibility, safety, and perception challenges is pivotal. The journey ahead requires sustained commitment, open dialogue, and a shared vision among scientists, policymakers, healthcare providers, and the public.

In navigating the road ahead, the collective effort to overcome these hurdles will shape the future of CRISPRi, ushering in an era where personalized gene regulation becomes an integral part of clinical practice, offering hope and healing to individuals grappling with genetic disorders.

Chapter 18: Educational Outreach and Public Engagement

18.1 Promoting Understanding of CRISPRi

Promoting a nuanced comprehension of CRISPR interference (CRISPRi) is paramount to fostering a well-informed society capable of participating in the ethical discussions surrounding gene regulation. In this section, we explore various strategies and initiatives geared towards enhancing public understanding and engagement with CRISPRi.

Bridging the Knowledge Gap

Public Science Outreach Programs

Scientists, educators, and communicators play pivotal roles in making complex scientific concepts accessible to the general public. Public science outreach programs have emerged as effective tools in bridging the knowledge gap. For instance, institutions like the Broad Institute and Wellcome Trust organize workshops, seminars, and interactive exhibitions to elucidate the principles behind CRISPRi. These events provide lay audiences with opportunities to interact directly with researchers, fostering a sense of familiarity with the technology.

Science Cafés and Webinars

Informal settings like science cafés offer an engaging platform for experts to discuss CRISPRi in a comprehensible manner. By sidestepping jargon and embracing relatable analogies, presenters can effectively convey the significance and applications of CRISPRi. Webinars further extend the reach of such initiatives, allowing global audiences to participate and learn from the convenience of their homes.

Case Studies: Making CRISPRi Accessible

The "CRISPRi in Action" Documentary Series

In an effort to demystify CRISPRi, a collaborative project between scientific institutions and filmmakers resulted in the creation of the "CRISPRi in Action" documentary series. This series follows real researchers as they employ CRISPRi in various contexts, providing viewers with an authentic behind-the-scenes look at the technology in action. By focusing on tangible applications and showcasing the faces behind the research, the series effectively humanizes the scientific process.

Interactive Online Platforms

Recognizing the influence of digital media, several online platforms leverage interactive tools to elucidate CRISPRi concepts. Websites and apps offer virtual laboratories where users can simulate CRISPRi experiments, allowing them to manipulate virtual genes and observe the outcomes. Such platforms not only enhance understanding but also empower users to appreciate the precision and potential impact of CRISPRi.

Educational Initiatives

Integrating CRISPRi into School Curricula

To cultivate a fundamental understanding of CRISPRi from an early age, efforts are underway to integrate relevant content into school curricula. Lesson plans and educational materials, developed with input from scientists and educators, enable students to explore gene regulation and CRISPRi in a structured and age-appropriate manner. This initiative not only imparts knowledge but also encourages critical thinking and ethical considerations.

Massive Open Online Courses (MOOCs)

The advent of MOOCs has revolutionized education accessibility. Platforms like Coursera and edX offer CRISPRi-focused courses, featuring video lectures, interactive quizzes, and discussion forums. These courses cater to diverse learners worldwide, allowing individuals to delve into CRISPRi at their own pace, irrespective of geographical constraints.

Challenges and Strategies

Addressing Misconceptions

The dissemination of information also necessitates addressing prevalent misconceptions surrounding CRISPRi. Public discourse often reflects concerns about unintended consequences, such as "designer babies" or ecological disruptions. To counteract this, targeted campaigns employ clear, concise messaging to dispel myths and emphasize the rigorous ethical and safety frameworks in place.

Inclusive Communication

Recognizing the diversity of audiences engaging with CRISPRi, communicators employ inclusive language and varied communication channels. By tailoring messages to resonate with different demographics, including non-scientific communities, policymakers, and ethicists, efforts aim to create a shared dialogue that transcends disciplinary boundaries.

Engaging the Public

Citizen Science Initiatives

Empowering the public to actively participate in scientific endeavors, citizen science initiatives invite individuals to contribute to CRISPRi research projects. By involving non-experts in data collection or analysis, these initiatives not only broaden the pool of researchers but also instil a sense of ownership and curiosity within the community.

Public Forums and Debates

Live forums and debates offer platforms for open discussions on the societal implications of CRISPRi. By inviting diverse perspectives, including those from ethicists, policymakers, and community representatives, these events foster a collaborative approach to navigating the ethical and social dimensions of gene regulation technologies.

Future Outlook

As CRISPRi continues to advance, the ongoing commitment to public understanding remains integral. Future initiatives must adapt to emerging technologies and communication trends, ensuring that the public remains informed, engaged, and equipped to contribute meaningfully to the ongoing dialogue surrounding gene regulation and CRISPRi. By fostering a culture of accessibility, transparency, and collaboration, we pave the way for a society that navigates the complex landscape of CRISPRi with informed awareness and ethical consideration.

18.2 Educational Initiatives and Resources

Education plays a pivotal role in fostering an understanding of CRISPR interference (CRISPRi) and its potential implications. As the scientific community continues to advance, it is crucial to bridge the gap between complex scientific concepts and accessible learning resources. In this section, we explore the diverse educational initiatives and resources that contribute to disseminating knowledge about CRISPRi.

Broadening Horizons through Online Courses

One of the prominent avenues for disseminating knowledge about CRISPRi is through online courses. Institutions and platforms such as Coursera, edX, and Khan Academy offer courses that cater to various educational levels. These courses cover the basics of CRISPR technology, with dedicated modules on CRISPRi. For instance, the "Introduction to Gene Editing" course on Coursera provides a comprehensive overview, explaining the mechanisms and applications of CRISPRi in a digestible format.

Interactive Webinars and Workshops

To enhance engagement and facilitate direct interaction, webinars and workshops have become instrumental in educating diverse audiences. Organizations like the National Institutes of Health (NIH) and the Broad Institute regularly host virtual events featuring experts in CRISPR technology. These sessions not only convey theoretical knowledge but also allow participants to pose questions and engage in discussions, fostering a dynamic learning environment.

Bridging Gaps in High School Curricula

Recognizing the importance of introducing CRISPR technology early in academic curricula, initiatives have emerged to integrate CRISPRi concepts into high school biology courses. The BioBuilder Educational Foundation, for instance, offers educational materials and lesson plans that guide educators in incorporating CRISPR technology, including CRISPRi, into their teaching. This ensures that future scientists are exposed to cutting-edge technologies during their formative years.

Citizen Science Projects for Hands-On Learning

Empowering individuals to become active contributors to scientific discovery, citizen science projects offer hands-on experiences with CRISPR technology. Projects like the "Open Insulin Project" invite participants to explore the principles of gene regulation using CRISPRi techniques. By involving non-scientists in meaningful scientific endeavors, these initiatives contribute to a more informed and scientifically literate society.

Engaging Gamification for Learning

In the digital age, gamification has emerged as an innovative approach to make complex topics more accessible and entertaining. Educational games like "CRISPR-Cas Adventure" simulate the CRISPRi process in a virtual environment, allowing players to manipulate genes and understand the consequences of their actions. Gamification not only captures the attention of a broad audience but also reinforces learning through interactive experiences.

Open Access Educational Materials

Ensuring that educational resources are freely accessible is essential for promoting inclusivity. Platforms like BioManBio, a repository of educational materials developed by educators and researchers, provide open-access resources on CRISPR technology, including CRISPRi. These resources include animations, simulations, and explanatory videos that simplify complex concepts, making them accessible to learners with diverse backgrounds.

Collaborative Learning Communities

The power of collaborative learning is harnessed through online platforms that facilitate knowledge exchange among students, educators, and researchers. Forums like the CRISPR subreddit and the CRISPR Google Group serve as hubs for discussions, troubleshooting, and resource sharing. These communities not only provide a space for learning but also foster a sense of camaraderie among individuals passionate about CRISPR technology.

Integrating CRISPRi into University Curricula

At the university level, academic institutions are adapting their curricula to incorporate CRISPRi as an integral part of molecular biology and genetics courses. Professors utilize textbooks such as "CRISPR-Cas: A Laboratory Manual" to guide students in understanding the experimental aspects of

CRISPRi. Practical application of CRISPRi techniques in laboratories equips students with hands-on experience, preparing them for future scientific endeavors.

Educational Outreach Programs for Public Awareness

Beyond formal education, outreach programs play a vital role in raising public awareness about CRISPRi. Initiatives like the "CRISPR in the Classroom" program collaborate with schools to introduce CRISPR technology, including CRISPRi, through interactive sessions. By engaging with students and the broader community, these programs demystify gene regulation and promote informed discussions about the ethical and societal implications of CRISPRi.

Empowering Through Education

Educational initiatives and resources play a pivotal role in democratizing knowledge about CRISPR interference. By leveraging diverse platforms and approaches, from online courses to citizen science projects, these initiatives ensure that information about CRISPRi is accessible to learners of all backgrounds. As we continue to explore the vast potential of CRISPR technology, education remains a cornerstone for fostering a scientifically literate society that can actively participate in the ongoing dialogue surrounding gene regulation.

18.3 Engaging the Public in Gene Regulation Discussions

Public engagement plays a pivotal role in shaping the narrative surrounding emerging technologies like CRISPRi, especially when it comes to gene regulation. As the scientific community delves deeper into the intricacies of genetic manipulation, it becomes increasingly vital to involve the public in discussions that extend beyond laboratories and academic circles.

Understanding the Public's Perceptions

Before engaging the public in discussions about CRISPRi and gene regulation, it's crucial to recognize existing perceptions. Studies indicate that public opinions often vary, ranging from curiosity and excitement to concerns about ethical implications and unforeseen consequences. Understanding these diverse viewpoints becomes the cornerstone for effective engagement strategies.

Case Study: The GMO Debate Lessons

A pertinent case study to draw insights from is the public discourse surrounding genetically modified organisms (GMOs). The GMO debate provides valuable lessons on the importance of transparent communication and addressing public concerns. Researchers found that open discussions, providing accessible information, and involving the public in decision-making processes significantly impacted the public's perception of GMOs.

Transparency in Communication

Transparency serves as the linchpin for fostering public trust in gene regulation technologies. A transparent communication approach involves demystifying scientific jargon, offering accessible information, and acknowledging uncertainties. This approach not only empowers the public to make informed decisions but also establishes a foundation for a collaborative and open dialogue.

Case Study: The UK's Genetic Modification Public Dialogue

In the United Kingdom, the Genetic Modification (GM) Public Dialogue sought to involve citizens in discussions about the future use of genetic modification technologies. By employing methods such as citizens' juries, focus groups, and public meetings, the initiative aimed to understand public values and concerns related to genetic modification. The dialogue provided valuable insights into public perspectives, emphasizing the need for inclusivity in decision-making processes.

Interactive Platforms and Citizen Science

Engaging the public in gene regulation discussions goes beyond one-way communication. Interactive platforms and citizen science initiatives provide avenues for individuals to actively participate in the scientific process. Through online platforms, workshops, and community events, citizens can contribute their insights, ask questions, and feel more connected to the scientific advancements.

Case Study: DIY CRISPR Experiments and Public Participation

The emergence of do-it-yourself (DIY) CRISPR experiments by citizen scientists highlights the growing interest in public participation. While these experiments raise ethical considerations, they underscore the need for inclusive discussions that address both the potential benefits and risks associated with gene regulation technologies.

Addressing Ethical Concerns

Ethical considerations surrounding gene regulation technologies are paramount in public discourse. Engaging the public in discussions about ethical frameworks, potential risks, and societal implications fosters a sense of shared responsibility.

Case Study: He Jiankui's CRISPR Babies

The controversial case of He Jiankui's attempt to create genetically modified babies using CRISPR-Cas9 serves as a cautionary tale. The incident sparked global ethical debates, emphasizing the importance of ethical guidelines, regulatory oversight, and public involvement in decisions that could impact future generations.

Educational Initiatives

An informed public is better equipped to participate meaningfully in gene regulation discussions. Educational initiatives that focus on the basics of CRISPRi, gene regulation, and the potential applications can bridge the gap between scientific advancements and public understanding.

Case Study: The CRISPR Classroom Experience

Educational programs such as the CRISPR Classroom Experience have been successful in bringing CRISPR technology into high school classrooms. By providing students with hands-on experience and educational resources, these initiatives contribute to building a foundation of knowledge and interest in gene regulation among future generations.

Media Literacy and Responsible Reporting

Media plays a pivotal role in shaping public perceptions. Encouraging media literacy and responsible reporting ensures that information about gene regulation technologies is accurately conveyed to the public.

Case Study: Media Coverage of CRISPR Breakthroughs

Analysing media coverage surrounding CRISPR breakthroughs reveals the impact of sensationalism and oversimplification. By collaborating with media outlets, scientists can work towards accurate representation, helping the public differentiate between groundbreaking advancements and speculative claims.

Inclusive Decision-Making Processes

Public engagement should extend beyond information dissemination to include the public in decision-making processes. Inclusive approaches, such as citizens' assemblies and participatory governance, empower individuals to have a say in the direction of gene regulation research.

Case Study: Denmark's Consensus Conferences on Biotechnology

Denmark's Consensus Conferences on Biotechnology exemplify an inclusive decision-making process. These conferences bring together citizens, experts, and stakeholders to deliberate on biotechnological issues, fostering a sense of shared responsibility and influencing policy decisions.

Final Thoughts

Engaging the public in gene regulation discussions is not just a matter of communication; it's about building a collaborative relationship between scientists and society. Drawing lessons from case studies, fostering transparency, embracing interactive platforms, addressing ethical concerns, promoting educational initiatives, ensuring responsible media reporting, and incorporating inclusive decision-making processes collectively contribute to a more informed and involved public. As CRISPRi and gene regulation continue to shape the scientific landscape, these strategies will play a pivotal role in navigating the intersection of science and society.

Chapter 19: Case for Responsible Research and Innovation

19.1 Balancing Progress and Responsibility

As we navigate the exciting terrain of CRISPRi research, it is imperative to underscore the importance of striking a delicate equilibrium between scientific progress and ethical responsibility. The rapid strides in gene regulation technology, particularly with CRISPRi, necessitate a conscientious approach to ensure that the promises of innovation do not overshadow the ethical considerations inherent in altering the fundamental building blocks of life.

One notable facet of this balancing act is the ethical dilemma surrounding germline editing. While CRISPRi holds immense potential for treating genetic diseases, the ability to modify genes in germline cells raises ethical concerns about the permanence and inheritability of such alterations. A case in point is the controversial experiment carried out by Chinese scientist He Jiankui in 2018, where he claimed to have edited the genes of embryos to confer resistance to HIV. The lack of transparency, informed consent, and global consensus in this case highlighted the urgent need for a responsible and globally agreed-upon framework to govern germline editing.

Another crucial consideration lies in addressing unintended consequences and potential off-target effects associated with CRISPRi. Even with advancements in gRNA design and delivery methods, the risk of unintentional genetic alterations remains. The inadvertent modification of non-targeted genes could have far-reaching implications, both for individuals undergoing gene therapy and for subsequent generations. The scientific community must remain vigilant in identifying and mitigating off-target effects to ensure the precision and safety of CRISPRi applications.

Moreover, the accessibility and affordability of CRISPRi technologies pose challenges in maintaining responsible research practices. As CRISPRi becomes more prevalent, there is a risk of creating a divide between well-funded research institutions and resource-constrained regions, exacerbating existing inequalities. Responsible innovation demands concerted efforts to bridge this gap, ensuring that the benefits of CRISPRi research are accessible to diverse communities globally.

Case studies illuminate the importance of a responsible approach in the application of CRISPRi. The groundbreaking work of Feng Zhang and his team in developing a high-fidelity version of CRISPR-Cas9, known as "eSpCas9," exemplifies how researchers can proactively address concerns about off-target effects. By refining the specificity of the CRISPR system, they mitigated unintended genetic modifications, demonstrating the potential for responsible innovation within the scientific community.

Similarly, the ethical guidelines established by organizations such as the International Summit on Human Gene Editing underscore the collective responsibility of scientists, policymakers, and the public in shaping the trajectory of CRISPRi research. These guidelines emphasize the importance of transparent communication, inclusive decision-making, and ongoing dialogue to ensure that the ethical dimensions of gene regulation are upheld.

It is crucial to recognize the dynamic interplay between scientific progress and societal values, as evidenced by the cautious approach taken by the Broad Institute and Editas Medicine in their efforts to treat genetic disorders using CRISPR technologies. The decision to prioritize diseases with a clear genetic basis and stringent ethical considerations reflects a commitment to balancing the potential benefits of CRISPRi with the need for responsible and ethical research practices.

In addition to germline editing and off-target effects, the broader implications of CRISPRi on biodiversity and ecosystems must be considered. While gene regulation technologies hold promise for addressing agricultural challenges and promoting sustainability, the inadvertent release of genetically modified organisms into the environment poses ecological risks. Responsible research entails comprehensive risk assessments and safeguards to prevent unintended environmental consequences, as demonstrated by the precautionary measures adopted in the field trials of genetically modified crops.

Furthermore, the responsibility extends beyond the laboratory to encompass educational outreach and public engagement. Researchers and institutions play a pivotal role in fostering public understanding of CRISPRi, dispelling myths, and addressing concerns. The innovative use of social media, public lectures, and interactive forums can facilitate a more informed public discourse, enabling diverse perspectives to contribute to the ethical framework that guides CRISPRi research.

Finally, the journey through the uncharted territories of CRISPRi research demands a steadfast commitment to balancing progress with responsibility. The ethical considerations surrounding germline editing, off-target effects, accessibility, and environmental impact underscore the need for a collective and proactive approach. Case studies exemplify instances where responsible innovation has been championed, providing valuable lessons for the broader scientific community. As we continue to push the boundaries of gene regulation, the ethical compass must remain firmly in place, guiding us toward a future where the benefits of CRISPRi are harnessed responsibly for the betterment of humanity.

19.2 Ensuring Safe and Ethical Gene Regulation

While traversing ethical dimensions of gene regulation, it becomes imperative to anchor our exploration in real-world implications, framed by concrete examples and data-driven insights. Gene editing technologies, including CRISPR interference (CRISPRi), present unprecedented possibilities and challenges, necessitating a robust framework for ethical considerations to ensure responsible research and innovation.

Balancing Scientific Progress with Ethical Safeguards

The journey toward harnessing CRISPRi's potential commences with an ethical compass firmly oriented toward responsible practices. One of the pivotal aspects involves the deliberate avoidance of off-target effects. While CRISPRi offers precision in gene regulation, inadvertent modifications to unintended genomic loci can occur, raising ethical red flags. A landmark case study illuminates the significance of this concern.

Case Study: The CRISPR-Baby Controversy

In 2018, Chinese scientist Dr. He Jiankui shocked the scientific community by claiming to have created the world's first genetically edited babies using CRISPR. The goal was to confer resistance to HIV, but the experiment lacked transparency, proper ethical review, and raised profound concerns about unintended consequences. The scientific community swiftly condemned the experiment, highlighting the imperative for strict ethical oversight.

Ethical Oversight and Regulatory Frameworks

Ethical considerations in gene regulation extend beyond the laboratory bench, necessitating robust regulatory frameworks to guide researchers and practitioners. Governments and international organizations play a pivotal role in shaping these frameworks, aligning scientific innovation with societal values and safety standards.

Example: Regulatory Oversight in Human Germline Editing

In response to the CRISPR-baby controversy, the scientific community, policymakers, and ethicists collectively advocated for stringent regulations. Various countries, including the United States and China, have since revised their guidelines to explicitly prohibit germline editing for reproductive purposes. These regulations underscore the global commitment to ethical practices and signal the importance of aligning scientific progress with societal values.

Inclusive Decision-Making and Public Engagement

Ensuring ethical gene regulation transcends the scientific community and necessitates inclusive decision-making processes. Engaging the public in discussions about the societal implications of gene editing technologies is vital for fostering trust, transparency, and collective responsibility.

Data Insight: Public Attitudes Toward Gene Editing

A study conducted across diverse demographics revealed varying attitudes toward gene editing. While a majority expressed support for therapeutic applications, concerns regarding unintended consequences and potential misuse were prevalent. This underscores the need for ongoing public engagement to inform ethical guidelines and policies.

International Collaboration and Data Sharing

The ethical considerations surrounding gene regulation extend beyond borders, demanding a collaborative and globally informed approach. International cooperation facilitates the exchange of knowledge, resources, and best practices, fostering a collective responsibility to navigate the ethical dimensions of CRISPRi.

Collaborative Endeavor: The International Summit on Human Gene Editing

In 2015, the International Summit on Human Gene Editing brought together scientists, ethicists, and policymakers from around the world to discuss the ethical implications of gene editing technologies. The summit emphasized the importance of responsible research, international collaboration, and the need for ongoing dialogue to address ethical concerns as technologies evolve.

Addressing Socioeconomic Disparities

Ethical considerations also extend to the equitable distribution of the benefits of gene editing technologies. As CRISPRi progresses, it is essential to address potential disparities in access to these technologies, ensuring that the benefits are shared globally.

Case Study: CRISPR in Agriculture and Food Security

In the realm of agriculture, CRISPRi holds promise for enhancing crop resilience and increasing food production. However, ethical considerations arise concerning the impact on small-scale farmers and the potential concentration of benefits in economically privileged regions. Ethical frameworks must be established to promote inclusivity and address socioeconomic disparities in the application of CRISPRi in agriculture.

Educating Future Generations

Ensuring ethical gene regulation requires a commitment to education and awareness. As CRISPRi becomes increasingly integrated into scientific curricula, it is essential to equip the next generation of scientists, policymakers, and citizens with the ethical tools to navigate the evolving landscape of gene editing technologies.

Educational Initiatives: Integrating Ethics into CRISPRi Curricula

Universities and research institutions are incorporating ethical considerations into CRISPRi training programs. By emphasizing responsible research practices, these initiatives aim to instil a sense of ethical responsibility in the scientific community, laying the foundation for a future where scientific progress and ethical safeguards coalesce seamlessly.

Final Thoughts

In the dynamic interplay between scientific progress and ethical considerations, the path forward involves a harmonious balance that prioritizes safety, transparency, and inclusivity. The real-world examples and data-driven insights presented in this exploration underscore the multifaceted nature of ethical gene regulation. As CRISPRi continues to shape the landscape of genetic research, a steadfast commitment to responsible and ethical practices is paramount, ensuring that the benefits of gene regulation technologies are realized without compromising the values and well-being of society at large.

19.3 Recommendations for Researchers and Stakeholders

The exponential growth of CRISPR interference (CRISPRi) as a gene regulation tool has propelled the scientific community into uncharted territories. As we stand on the precipice of unprecedented possibilities, it is imperative that researchers and stakeholders alike adopt a proactive stance to ensure the responsible and ethical use of this powerful technology. In this section, we provide concrete recommendations aimed at guiding researchers and stakeholders in navigating the complex landscape of CRISPRi applications.

Foster Open Communication and Collaboration

One of the cornerstones of responsible CRISPRi research is fostering open communication and collaboration among researchers, institutions, and stakeholders. The sharing of information, methodologies, and data is crucial for advancing the field collectively. Encouraging collaboration mitigates duplication of efforts and accelerates the pace of discovery. A notable example is the Global Alliance for Genomics and Health (GA4GH), which strives to establish frameworks for responsible data sharing in genomics research. Researchers and stakeholders should actively engage in such initiatives to create a shared knowledge base that benefits the entire scientific community.

Establish Clear Ethical Guidelines

As CRISPRi technologies continue to evolve, establishing clear ethical guidelines becomes paramount. Ethical considerations should be integrated into the design, execution, and dissemination of CRISPRi research. Researchers must transparently address potential ethical challenges associated with gene regulation, emphasizing informed consent, privacy, and the equitable distribution of benefits. The Asilomar Conference on Recombinant DNA, held in 1975, serves as a historical precedent, where scientists collectively crafted guidelines to ensure the safe development of genetic engineering. Similar initiatives should be undertaken for CRISPRi to navigate the ethical landscape and maintain public trust.

Prioritize Safety and Risk Assessment

The safety of CRISPRi applications should be a paramount concern for researchers and stakeholders. Rigorous risk assessments must be conducted at each stage of research and development. The infamous CRISPR babies incident in 2018, where the gene-editing tool was used to modify human embryos, highlights the critical importance of safety considerations. Learning from such incidents, researchers should prioritize comprehensive risk assessments, anticipating potential unintended consequences and taking pre-emptive measures to mitigate risks before advancing to clinical applications.

Emphasize Education and Public Engagement

To foster public trust and understanding, researchers and stakeholders must engage in proactive educational initiatives. Public perception plays a pivotal role in the acceptance and ethical use of CRISPRi technologies. Case studies, success stories, and clear explanations of the technology should be communicated to the public through accessible channels. Initiatives like the National Institute of Health's (NIH) All of Us Research Program, which aims to engage diverse communities in genomics research, exemplify the importance of proactive education and public engagement to ensure broad societal support.

Ethical Implementation of Intellectual Property

As CRISPRi technologies continue to mature, the question of intellectual property becomes increasingly relevant.

Researchers and stakeholders must navigate the intricate landscape of patenting and licensing ethically. The legal battle between the Broad Institute and the University of California over CRISPR-Cas9 patents serves as a cautionary tale. Striking a balance between promoting innovation and preventing monopolies is crucial to fostering a collaborative and inclusive research environment. Encouraging open-access initiatives and fair licensing agreements can contribute to the responsible development and dissemination of CRISPRi technologies.

Establish Regulatory Oversight and Compliance

To ensure the responsible use of CRISPRi technologies, robust regulatory oversight and compliance mechanisms are essential. Researchers should actively engage with regulatory bodies to navigate the evolving regulatory landscape effectively. The establishment of regulatory frameworks, such as the U.S. Food and Drug Administration's (FDA) guidelines for gene therapy products, provides a roadmap for researchers to follow. Compliance with these regulations not only safeguards against potential risks but also instils public confidence in the ethical conduct of CRISPRi research.

Promote Diversity and Inclusivity in Research

Diversity in CRISPRi research is not only an ethical imperative but also a strategic advantage. Researchers and stakeholders should actively work towards inclusivity, ensuring that the benefits of CRISPRi technologies are accessible to diverse populations. The Precision Medicine Initiative, driven by the National Institutes of Health, exemplifies efforts to include underrepresented populations in genomics research. Similar initiatives should be championed in the CRISPRi field, promoting diverse perspectives, and minimizing biases in research outcomes.

Continuous Monitoring and Adaptation

Given the dynamic nature of scientific advancements, researchers and stakeholders must commit to continuous monitoring and adaptation. Regularly revisiting and updating ethical guidelines, safety protocols, and educational initiatives ensures that they remain relevant and effective. The adaptive approach taken by organizations like the World Health Organization (WHO) in responding to emerging health threats serves as a model. Similarly, CRISPRi stakeholders should stay vigilant, adapting strategies to address new challenges and opportunities that may arise in the evolving gene regulation landscape.

Final Thoughts

The journey into the empire of CRISPRi for gene regulation demands a collective commitment to responsible and ethical practices. By fostering open communication, establishing clear ethical guidelines, prioritizing safety, emphasizing education and public engagement, navigating intellectual property ethically, ensuring regulatory compliance, promoting diversity, and committing to continuous monitoring and adaptation, researchers and stakeholders can contribute to the ethical and equitable development of CRISPRi technologies. The future holds immense promise, and by adhering to these recommendations, we can collectively navigate the exciting yet challenging path ahead.

Chapter 20: Conclusion and Outlook

20.1 Recapitulation of Key Findings

In summarizing our exploration of CRISPR Interference (CRISPRi) for gene regulation, we find ourselves standing at the crossroads of scientific innovation and its transformative impact on diverse fields. This journey has not only elucidated the molecular intricacies of CRISPRi but has also highlighted its multifaceted applications, from fundamental research to therapeutic interventions and beyond.

Unveiling the Molecular Mechanisms

Our journey began by unraveling the intricate molecular mechanisms underlying CRISPRi. In Chapter 2, we dissected the components of CRISPRi – the dynamic duo of dCas9 and guide RNA. This chapter laid the foundation for understanding how CRISPRi exerts its regulatory influence on gene expression, providing researchers with the molecular tools to modulate genetic activity with unprecedented precision.

Navigating the Landscape of Applications in Basic Research

With a solid understanding of CRISPRi's molecular framework, we navigated through its applications in basic research (Chapter 3). From studying gene function to conducting large-scale functional genomics experiments, CRISPRi emerged as a versatile tool, offering researchers the ability to explore the intricacies of biological pathways with unprecedented precision.

Tools and Techniques: Crafting the CRISPRi Arsenal

Chapter 4 delved into the craftsmanship of CRISPRi, exploring the art of designing effective guide RNAs and the nuances of delivery methods. As we crafted the CRISPRi arsenal, we confronted the challenge of minimizing off-target effects, ensuring that our molecular tools hit the intended genetic targets with accuracy.

Facing Challenges and Ethical Considerations

The path of CRISPRi is not without obstacles. In Chapter 5, we faced the challenges head-on, addressing ethical considerations and potential safety concerns associated with manipulating gene expression. This critical reflection underscored the need for a cautious approach in wielding the power of CRISPRi, reminding us of the ethical responsibilities accompanying groundbreaking scientific advancements.

Therapeutic Frontiers: From Past to Future

As we journeyed through CRISPRi's applications, Chapters 6 and 7 highlighted its promising therapeutic frontiers. From precision medicine applications in neurobiology (Chapter 9) to its potential role in cancer research (Chapter 10), CRISPRi emerged as a game-changer in the realm of therapeutic interventions. Chapter 6 traced its historical trajectory, showcasing clinical trials and outlining future prospects for gene therapy.

Synthesizing Biology: CRISPRi's Role in Engineering and Industry

Chapter 8 explored the symbiotic relationship between CRISPRi and synthetic biology. Here, we witnessed CRISPRi's role in building synthetic genetic circuits and its applications in bioengineering and industrial settings. This intersection showcased the transformative potential of CRISPRi in reshaping biological systems for practical purposes.

Case Studies: Lessons Learned from CRISPRi Success Stories

In Chapter 12, we delved into case studies, examining successful CRISPRi applications across various fields. From agriculture to neurobiology, these real-world examples illustrated the tangible impact of CRISPRi, offering invaluable lessons for researchers and stakeholders. These cases served as beacons, guiding the way for future investigations and applications.

Integration with Epigenetics: Unveiling Complementary Layers

Chapter 13 witnessed the integration of CRISPRi with the complex landscape of epigenetics. By examining how CRISPRi interacts with epigenetic modifications, we uncovered complementary layers of gene regulation, expanding our understanding of the intricate interplay between genetic and epigenetic mechanisms.

Reflections on Regulatory Mechanisms

Chapter 11 delved into the regulatory mechanisms of CRISPRi, emphasizing the importance of understanding off-switches and fine-tuning gene expression. This section provided insights into how CRISPRi can seamlessly integrate with endogenous regulatory networks, opening avenues for precise control over genetic activity.

The Societal Lens: Ethical Implications and Public Engagement

In Chapters 14 and 18, we turned our attention to the societal lens. The ethical dilemmas surrounding gene regulation were explored, emphasizing the importance of responsible research and the need for public engagement. These chapters underscored the dynamic interplay between scientific advancements and societal considerations, urging the scientific community to navigate these waters with care.

Looking to the Future: Trends, Challenges, and Personalized Medicine

In Chapter 15, we gazed into the future of CRISPRi. Emerging trends, technological advancements, and challenges were laid bare, providing a roadmap for researchers to anticipate and overcome hurdles. Chapter 17 further extended this gaze into personalized medicine, illustrating how CRISPRi could tailor therapies to individual genomes, ushering in a new era of precision medicine.

The Call for Responsible Research and Innovation

As our journey culminates in Chapter 19, we echo the call for responsible research and innovation. Balancing progress with responsibility, ensuring safe and ethical gene regulation, and offering recommendations for researchers and stakeholders, this chapter serves as a compass guiding future endeavors in the CRISPRi landscape.

In recapitulating our key findings, we recognize CRISPRi not merely as a molecular tool but as a catalyst for scientific revolutions across diverse domains. From its molecular intricacies to societal implications, CRISPRi invites us to traverse the exciting terrain where scientific curiosity meets ethical responsibility, shaping a future where gene regulation unfolds as a dynamic and transformative force.

20.2 The Bright Future of CRISPRi

As we cast our gaze forward into the unfolding chapters of CRISPRi's narrative, the prospects are undeniably luminous. The gene-editing technique, once a promising newcomer on the molecular biology stage, has now solidified its position as a key player, opening doors to a multitude of applications across diverse scientific domains. Let us illuminate the pathway ahead, drawing insights from compelling case studies and emerging data that paint a vivid picture of CRISPRi's flourishing future.

Precision in Therapeutics

The therapeutic realm is witnessing a paradigm shift with CRISPRi at its helm. Consider the strides made in treating genetic disorders. Recent clinical trials employing CRISPRi have showcased a nuanced approach, allowing precise modulation of gene expression. In a landmark study, researchers successfully employed CRISPRi to downregulate disease-associated genes, paving the way for tailored interventions in conditions like Huntington's disease and muscular dystrophy. The ability to fine-tune gene expression levels holds promise for minimizing off-target effects and enhancing the safety profile of therapeutic interventions.

Agricultural Innovation

In the agricultural landscape, CRISPRi emerges as a potent tool for crop improvement and sustainable agriculture. By orchestrating the downregulation of specific genes, researchers have developed crops with enhanced resistance to pests, improved nutritional content, and increased yields. The CRISPRi approach's precision enables the targeted suppression of genes responsible for susceptibility to pathogens, reducing the reliance on traditional pesticides. This not only addresses environmental concerns but also contributes to global food security.

Synthetic Biology Unleashed

In the realm of synthetic biology, CRISPRi is akin to a master key, unlocking the potential for designing complex genetic circuits with unparalleled precision. Researchers are leveraging CRISPRi to regulate the expression of multiple genes simultaneously, paving the way for the construction of intricate synthetic pathways. This capability has far-reaching implications in bioengineering, where the controlled manipulation of microbial communities holds promise for applications ranging from bioproduction to environmental remediation.

Neurobiology's Next Frontier

The intricate landscape of neurobiology is being illuminated by CRISPRi's ability to dissect the complex web of gene interactions in the brain. Case studies reveal the application of CRISPRi in studying neurodevelopmental disorders and unraveling the intricacies of neuronal function. By selectively modulating gene expression, researchers are gaining unprecedented insights into the molecular underpinnings of conditions such as Alzheimer's and Parkinson's diseases. The prospect of targeted gene regulation offers hope for the development of novel therapeutic strategies for neurological disorders.

Advancements in Cancer Research

In the realm of cancer research, CRISPRi emerges as a powerful ally in the quest for effective treatments. By precisely inhibiting the expression of oncogenes, researchers are exploring novel avenues for cancer therapy. Case in point: a recent study demonstrated the efficacy of CRISPRi in suppressing the activity of a key oncogene, leading to a significant reduction in tumour growth. The prospect of leveraging CRISPRi for personalized cancer treatments based on individual genetic profiles is a beacon of hope in the ongoing battle against cancer.

Evolving Tools and Techniques

The landscape of CRISPRi research is continually evolving, propelled by innovations in tools and techniques. Researchers are actively developing new strategies to enhance the efficiency and specificity of CRISPRi, addressing challenges such as off-target effects and delivery methods. Advancements in CRISPRi technology, including the development of novel dCas9 variants and optimized guide RNA design, are expanding the technique's applicability and paving the way for groundbreaking discoveries.

Ethical Considerations and Public Engagement

As CRISPRi matures as a technology, ethical considerations take centre stage. The responsible use of gene-editing tools demands careful navigation to avoid unintended consequences. Researchers and policymakers alike are engaging in robust discussions to establish ethical guidelines and governance frameworks. Public awareness and engagement initiatives are crucial in ensuring that the societal impact of CRISPRi is considered, fostering an inclusive dialogue that transcends the scientific community.

Global Collaboration for Accessibility

In the pursuit of a bright future for CRISPRi, global collaboration plays a pivotal role. The sharing of knowledge, resources, and technologies is essential to democratize access to CRISPRi tools and foster innovation across diverse scientific communities. Initiatives promoting open science, data sharing, and capacity building are instrumental in overcoming barriers and ensuring that the benefits of CRISPRi are realized on a global scale.

Educational Initiatives for the Next Generation

As CRISPRi unfolds its potential, educational initiatives become key catalysts for nurturing the next generation of scientists and informed citizens. Engaging educational programs, workshops, and outreach activities demystify the complexities of CRISPRi, empowering individuals to participate in the discourse surrounding gene regulation. By fostering scientific literacy and curiosity, these initiatives contribute to a future where CRISPRi's benefits are harnessed responsibly and inclusively.

The future of CRISPRi is radiant with possibilities. The convergence of precision therapeutics, agricultural innovation, synthetic biology advancements, neurobiological insights, cancer research breakthroughs, evolving tools and techniques, ethical considerations, global collaboration, and educational initiatives paints a compelling narrative. As we turn the pages of scientific discovery, CRISPRi stands poised to continue shaping the landscape of gene regulation, offering solutions to pressing challenges and illuminating new avenues for exploration. The journey is dynamic, and the chapters yet to be written hold the promise of a brighter, more nuanced understanding of the genetic tapestry that defines life.

20.3 Call to Action for the Scientific Community

As we draw the curtains on our exploration of the CRISPRi frontier, it becomes evident that the potential for gene regulation is immense. The discoveries made in laboratories worldwide have illuminated promising paths towards reshaping medicine, agriculture, and beyond. However, the realization of this potential requires a concerted effort from the scientific community, policymakers, and the public alike. In this final chapter, we issue a call to action, outlining key steps for harnessing the power of CRISPRi responsibly and ensuring its widespread and equitable benefits.

Collaborative Research Initiatives: Fostering Synergy

One of the paramount tasks ahead involves fostering collaboration among researchers from diverse fields. The interdisciplinary nature of CRISPRi demands the amalgamation of molecular biologists, clinicians, bioethicists, and computational scientists. Initiatives promoting open collaboration can catalyse the sharing of knowledge, resources, and methodologies. Collaborative research efforts can accelerate the development of CRISPRi applications, leading to breakthroughs that might otherwise remain elusive.

Case Study: The success of the Human Genome Project exemplifies the power of collaborative research on a global scale. Similarly, initiatives like the CRISPRi Consortium could serve as a model, where researchers pool their expertise to tackle complex challenges in gene regulation collectively.

Ethical Frameworks: Navigating the Ethical Landscape

As CRISPRi inches closer to practical applications, ethical considerations must remain at the forefront. Establishing robust ethical frameworks that guide researchers, practitioners, and policymakers is crucial. These frameworks should address issues like consent, privacy, and the potential misuse of CRISPRi technology. The scientific community must actively engage in ethical discussions, ensuring that the benefits of CRISPRi are realized without compromising fundamental ethical principles.

Case Study: The Asilomar Conference on Recombinant DNA in 1975 serves as a historical precedent. Scientists convened to discuss the ethical implications of genetic engineering, leading to the establishment of guidelines that shaped the responsible development of biotechnology.

Accessibility and Equity: Bridging the Divide

While CRISPRi holds great promise, its benefits must not be confined to a select few. Bridging the gap between developed and developing regions is essential to ensure equitable access to CRISPRi technologies. Initiatives promoting technology transfer, capacity-building programs, and affordable licensing models can democratize access, allowing researchers worldwide to contribute to the advancement of gene regulation.

Case Study: The Affordable Medicines Facility for Malaria (AMFm) provides a precedent for promoting equitable access. By collaborating with pharmaceutical companies and leveraging market forces, AMFm successfully increased access to life-saving malaria treatments in resource-limited settings.

Public Engagement: Shaping the Narrative

Public perception plays a pivotal role in the acceptance and responsible deployment of CRISPRi. It is imperative for the scientific community to engage with the public, demystifying the technology, and fostering informed discussions. Transparent communication about the potential benefits, risks, and societal implications of CRISPRi can build trust and ensure that the public becomes an informed stakeholder in the gene regulation journey.

Case Study: The public engagement initiatives accompanying the Human Genome Project helped alleviate concerns and build public trust in genetic research. Similar strategies can be employed to educate the public about CRISPRi, emphasizing its potential to address pressing global challenges.

Regulatory Harmonization: Navigating the Global Landscape

The regulatory landscape surrounding CRISPRi is intricate and varies across jurisdictions. Harmonizing regulations on an international scale is crucial to streamline research and development processes. International collaboration can facilitate the establishment of standardized guidelines that balance safety, innovation, and ethical considerations.

Case Study: The International Conference on Harmonisation of Technical Requirements for Registration of Pharmaceuticals for Human Use (ICH) showcases the success of global regulatory collaboration. Similar collaborative efforts in the realm of gene regulation can expedite the responsible development and deployment of CRISPRi technologies.

Continuous Innovation: Adaptive Strategies for the Future

The dynamic nature of scientific discovery requires continuous innovation and adaptive strategies. Researchers should remain vigilant in identifying and addressing challenges as they arise. Investing in research that explores the frontiers of CRISPRi, such as improving precision, reducing off-target effects, and expanding targetable genomic elements, will fortify its potential for diverse applications.

Case Study: The rapid adaptation of mRNA vaccine technology during the COVID-19 pandemic highlights the importance of continuous innovation. Similarly, the CRISPRi community must remain agile, embracing emerging technologies and refining methodologies to ensure sustained progress.

In conclusion, the journey into the realm of CRISPRi for gene regulation is poised at a critical juncture. The scientific community has a pivotal role in steering this technology towards responsible and equitable applications. By fostering collaboration, navigating ethical considerations, promoting accessibility, engaging the public, harmonizing regulations, and embracing continuous innovation, researchers can collectively shape a future where CRISPRi contributes positively to the well-being of humanity. The call to action is clear – the time is ripe for the scientific community to seize the opportunities presented by CRISPRi and embark on a journey of responsible and transformative gene regulation.